Encyclopaedia of Life Science

Encyclopaedia of Life Science

VOL. 2

B.A. Mannan

ANMOL PUBLICATIONS PVT. LTD
NEW DELHI - 110 002 (INDIA)

ANMOL PUBLICATIONS PVT. LTD.
H.O.: 4374/4B, Ansari Road, Darya Ganj,
New Delhi-110 002 (India)
Ph.: 23278000, 23261597
B.O.: No. 1015, Ist Main Road, BSK IIIrd Stage
IIIrd Phase, IIIrd Block,
Bangalore - 560 085 (India)
Visit us at: www.anmolpublications.com

Encyclopaedia of Life Science

First Published, 2007

ISBN 978-81-261-3408-3 (Set)

PRINTED IN INDIA

Printed at Amit Enterprises Delhi.

Content

Preface

'Life Science' is a branch of science, under which we study about the living beings in detail. We study about the body structure; all the organs of the body and their functions. The body consists of several organs, which are made of Tissues, while, Tissues are made of cells. Plants are very useful for living organisms as they supply oxygen for the respiration and they are the food-source for the primary consumers. Plant survival and growth is critically influenced by abiotic factors including water, air and light. The effect of the local environment on plant growth also accounts for much of variation in adult form within a species. Plants have green pigments, which are quite useful in the process of photosynthesis. In the presence of light, plants make their food with the help of minerals, water and carbondioxide, in their green leaves, in the form of carbohydrates. The food prepared in green leaves is transported to different parts of the plant by phloem.

Similarly, animal life is a significant feature on earth. All the functions of the human beings and animals, including birds and insects, are performed in their cells. Cell is the structural unit of any living body. This was first discovered by scientist, Robert Hook. The cell has several parts, such as nucleus, centrosomes, mitochondria, plastids, cell-walls. The cell wall gives the mechanical support to the cell and cell-walls are accountable for the transportation of the material in and out of the cells.

Biologists agree that all living things arose through long periods of change shaped by physical and chemical processes. Approximately, there are two million species in the living world. In changing environment, a few species are eliminated during un-favourable conditions on the basis of the principle of the 'survival of the fittest,' which is also known as natural selection.

Science can strive to grasp nature's faculty, but science is socially embedded and all scientists record prevailing certainties. Darwin himself, in the closing lines of 'The Origin of Species', mentioned about Victorial social preference. Life's pathway certainly includes many features, predicable from laws of nature. But laws of nature do not tell us why we have crabs and snails at all, why insects rule the multicellular world, and why vertebrates rather than persistent algal mats exist as the most complex form of life on earth.

Bacteria too represent the great success story of life's pathway. They occupy a wider domain of environments and span a border range of biochemistries than any other group. One might conclude that complexification of life as a whole represents a pseudo-trend based on constraints at the wall, but still holds that evaluation continues, within a particular group.

This eminent work, titled; *Encyclopaedia of Life Science* would be able to make new waves in the discipline. Expectedly, these books would prove to be the best source of study in the area of Life Science.

Editor

1

Muscular System

Muscular movement is a characteristic of the higher animals. Movement in lower animals is just a primitive version of the muscular activities found in higher animals. In addition to locomotion, muscles perform a number of functions associated with digestion, respiration, circulation, excretion and reproduction. Thus, muscular activity plays a key role in animal life.

Classification of Muscles

The human body, like most higher animals, contains three distinct types of muscles: *smooth, skeletal* and *cardiac.*

(a) *Smooth Muscles:* These occur primarily in the wall of internal organs (*e.g.*, digestive tract, arteries, veins, etc.). They are spindle-shaped cells, each with a centrally located nucleus. Unlike skeletal and heart muscles, they are slow to contract, and do not have cross striations on their myofibrils. The contraction and relaxation of smooth muscles are directly controlled partly by the nervous system and partly by them. Unlike skeletal muscles, but like cardiac muscles, they are not under voluntary control.

(b) *Skeletal Muscles:* (also called Striated, Voluntary or Somatic muscles) contract most rapidly and vigorously. They are attached to the bones or tendons, so they are mainly responsible for the movements of the skeleton and therefore of the body. The tissue of skeletal muscle is formed by long cylindrical cells or fibres which are multinucleated. They have well developed cross striations. Each muscle fibre is composed of fibrils or myofibrils, having characteristically alternating light and dark regions. Many invertebrates, including even some coelenterates, also contain some striated muscles.

(c) *Cardiac or Heart Muscles*: These form a unique type of muscular tissue found only in the heart. They resemble with the skeletal muscles in being made up of striated multinucleated fibres. However, the nuclei of each fibre are not peripheral as in skeletal muscle, but are distributed within the cytoplasm. Cardiac muscle is functionally syncytial and contracts rhythmically in the absence of external innervation due to the presence in the myocardium of pacemaker cells that discharge spontaneously.

Fine Structure of the Skeletal Muscle Fibre: Each skeletal muscle fibre is an elongated multinucleate cell completely surrounded by a *sarcolemma,* the cell membrane. It consists of a semi-fluid cytoplasm called *sarcoplasm*, containing numerous longitudinal fibrils called myofibrils. The many nuclei scattered throughout the fibre are generally located immediately under the sarcolemma. The diameter of a muscle fibre ranges from 10 to 100 μ and they extend through the whole length of the muscle joining with the tendons at their ends. Sarcoplasm contains, in addition to several proteins and various enzymes, nuclei, numerous mitochondria, a small Golgi body, sarcoplasmic reticulum, lipid droplets, and glycogen. Because sarcosomes (mitochondria) are the sites of oxidativeone sarcomere within a myofibril energy metabolism, their presence in relatively large numbers in a muscle has usually been taken as an indication of a muscle that performs more work.

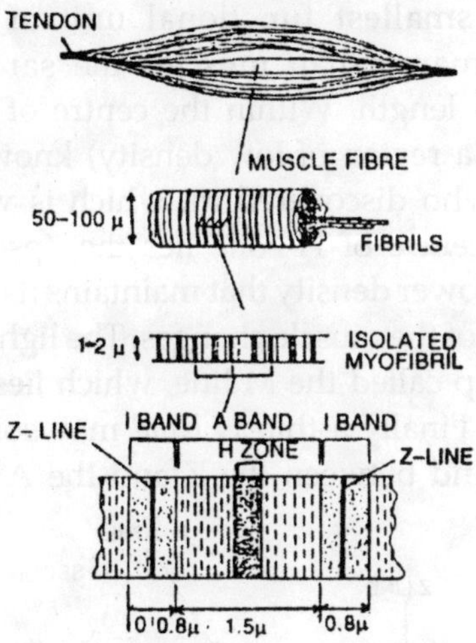

Schematic diagram of whole muscle fibre, fibril and banding pattern of one sarconere within a myofibril (Redawn from H. E. Huxley).

The *sarcaplasmic reticulum* is a system of tubules (sarcotubules) that does not open to the exterior. But the units of the other system, the *transverse tubules* open to the exterior of the fibre. These tubules are inwardly directed extensions of the sarcolemma and their cavities are thus in continuity with the fluid which surrounds the fibrils. Such continuity provides a simple means of conveying an electrical signal derived from the depolarisation of the sarcolemma.

Each *myofibril* has a diameter from about 1 to 2 microns. The myofibrils are the actual *contractile elements* (units) of the muscle fibre, and it is the coincidence of light and dark bands of many myofibrils, which gives the muscular fibre its characteristically striated appearance. The striations or repeating pattern of light and dark bands are similar for each myofibril and are designated by letters. The dense (dark) bands are called *A* or *anisotropic bands,* while the lighter bands are called I *(isotropic)* bands. In the middle of the I band can be recognised the Z-band or Z-line (from the German, *Zwischenscheibe,* central membrane) as a narrow dark line.

The area between two adjacent Z lines is called a *sarcomere,* which is the smallest functional unit of the muscle. In uncontracted mammalian muscles the sarcomere is about 2.1 to 2.4 μ in length. Within the centre of the A band is a lighter region (a region of low density) known as the H *zone* (after *Hensen* who discovered it), which is visible in relaxed fibrils. In the centre of H-zone lies the "pseudo H-zone," a region of even lower density that maintains its width no matter, how the length of the muscle changes. The light zone surrounds a dark thin strip called the M line, which lies in the centre of the sarcomere. Finally a thin N *band* may sometimes be seen within the I band between the Z and the A bands.

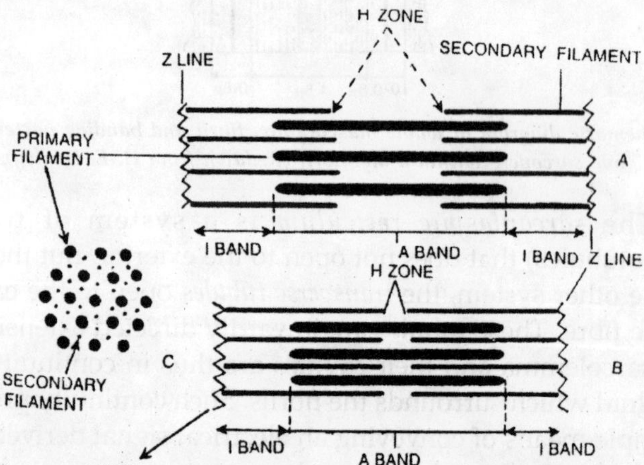

The arrangement of the primary and secondary filaments in myofibrils and the relationships of the two types of filaments in muscle contraction and relaxation. (A) Extended muscle, (B) Partially contracted muscle, (C) Cross-section through A band of myofibril (after H.E. Huxley).

Recent work of *H.E. Huxley* (1958,1965) has shown that each myofibril is composed of two types of filaments (*myofilaments*), known as the *primary* or *thick* filaments and the *secondary* or *thin filaments.* In vertebrate muscles, the thick filaments are about 100 A thick and 1.5 μ long while the thin or secondary filaments are approximately 50 Å thick and 2.0 μ long. Thus, each myofibril is constructed of a very large number of each type of filaments, arranged so that each primary filament is surrounded by 6 secondary filaments, which are in turn shared with 6 surrounding primary filaments as shown in Fig. The M line, described above, is thought to be caused by a slight bulge in the centre of each thick filament.

Hoyle and his coworkers, after examining a variety of invertebrate and vertebrate striated muscles, have suggested that a *superthin filament* (called *T-filament)* runs the length of many sacromeres. These filaments, 25 Å in diameter, were observed by these workers in all bands of the myofibril and were seen to bridge the gap between the actin and myosin filaments produced when a muscle fibre was stretched. *McNeil* and *Hoyle* (1967) and *Walcott* and *Ridgway* (1967) have suggested that superthin filaments serve as elastic supportive elements of the myofibril.

The repeating band pattern in the myofibrils is due to the alternate overlapping of thick and thin filaments, the dark bands occurring where both kinds of filaments overlap, and the light bands where only one type of filaments overlap.

The dense Z lines are continuous across the width of each muscle fibre and somehow serve to hold the myofibrils together. In this manner it accounts for the coincidence of A and I bands of the many myofibrils to give the whole muscle fibre its typically striated appearance. The two kinds of filaments are linked together by an intricate system of cross bridges which play an important role in muscle contraction.

Chemical Composition of the Muscle Fibre: Each muscle consists of at least five kinds of materials. Of which about 75 per cent is water, and nearly 20 per cent protein. The remaining 5 per cent includes glycogen, inorganic ions, and a variety of other organic substances like free amino acids and *phosphagens (phosphocreatine* in vertebrate muscles, and *phosphoarginine* of invertebrates). The phosphagens play an important role in muscle contraction. The muscle proteins are of 3 types.

1. *The Stroma Proteins:* They constitute about 1/5 of the total muscle proteins. They can be identified with collagen and elastin elements that bind muscle fibres together and transmit tension to tendons.
2. *The Ordinary Cellular Proteins:* They are not specifically characteristic to muscle cells and constitute another 1/5 of the total muscle proteins.

Most important of them are the enzymatic proteins which catalyse various chemical reactions of the muscle cells, and the proteins that supply energy to the contractile system.

3. *The Special Contractile Proteins:* The thick filaments are composed of the protein *myosin*. The thin filaments are made up of three different types of proteins—the globular proteins *actin* and *troponin,* and a long thin protein *tropomyosin*. One thin filament probably contains between 300 and 400 actin molecules, and about 50 tropomyosins and 50 troponins.

Proteins of the Myofibril

Protein	*Location*	*% of Myofibril Protein*	*MW*	*Chain Composition*
Myosin	Thick filament	43-45	521,000	2 heavy chains 4 light chains
Actin	Thin filament	21-23	41,900	1×41,900
Tropomyosin	In grooves of thin filament	5	70,000	2×35,000
Troponin	Thin filament at 400-Å intervals	5	70,500	1TN - T (1 × 30,500) 1TN-I (1× 22,000) 1 TN - C (1 × 18,000)
Actinin	Z-line	1	180,000	2×90,000
Titin*	I-band	8-10	2500,000 to 3000,000	1×2500,000 to 1 × 3000,000
Nebulin*	I-band	3-5	700,000	700,000
M-line protein	M-line	2	165,000	1×165,000
C - protein	Thick filament	1-2	140,000	1 × 140,000

**Recently discovered by K. Wang, Myosin, Actin and Tropomyosin are mechano-chemical proteins ; Troponyosin and Troponin are Regulatory proteins.*

Actin and myosin are absolutely essential for contraction. It is said that tropomyosin plays a vital role in sensitising the contractile proteins (actin and myosin) to calcium ions. This sensitivity to calcium is essential for switching the muscle contraction on and off.

Among the three proteins of a myofibril, myosin is the most abundant. Skeletal muscle contains twice as much myosin by weight as actin. By selective extraction procedures it has been possible to show that the thick filaments of the A-band are composed of myosin, while the thin filaments of the I-band are composed of actin and very likely tropomyosin. For example, after treatment of muscle with a suitable salt solution which is known to remove myosin, the A bands disappear. This is confirmed by the observation that similar treatment of myofibrils results in the removal of primary filaments. If actin of the myofibrils is extracted, most of the thin filaments and I bands are removed. The remaining fragments of I bands are possibly made up of tropomyosin.

(a) *Myosin:* Myosin molecules are tadpole-shaped with an overall length of 1520 ± 260 Å and a very high molecular weight about 521,000. Each molecule contains six polypeptide chains: *two heavy chains* (MW = 200,000) with identical sequences that account for most of the weight of the molecule, and *two pairs of light chains* that vary in size depending on the source (MW = 15,000 to 27,000). The NH_2 terminus of each heavy chain forms a *globular domain* to which the light chains are attached. This domain binds actin and contains the ATPase activity of myosin. The remaining part of each heavy chain is a fibrous, almost entirely α - helical structure that extends to the COOH terminus, coiled around the other heavy chain in the molecule.

Two of the myosin light chains can be dissociated from the globular domains at high pH, and are called *alkali light chains.* There are two different types of *alkali light chains:* α_{-1} (MW = 25,000) and A_2 (MW = 16,000). Both A chains have an identical COOH-terminal sequence of 141 residues preceded by an 8-residue segment that differs in five amino acid substitutions. The A_1 chain has an additional NH_2- terminal segment of 41 residues. Both A_1 and A_2 chains are required for myosin ATPase activity and binding to actin. The other pair of light chains (MW = 18,000), which can be dissociated from the globular domain by DNTB, 5,5'-dithiobis (2-nitrobenzoate) are designated DNTB *light chains* and have indentical sequences.

By brief digestion with trypsin, myosin molecules can be broken down into head and tail fragments. The head fragment is globular in shape measuring about 150-250 Å in length and 40 Å in diameter, with a molecular weight of about 350,000. This protein, also called *heavy meromyosin* (HMM), has two chemically different active sites. At one site binding to actin molecule and at the other hydrolysis of ATP occurs. The tail fragments of myosin molecules are called hydrolysis of ATP occurs. The tail fragments of myosin molecules are called *lightmeromyosin* (LMM).

They are about 1000 Å long, 15-20 Å in diameter, and with a molecular weight of about 150,000. LMM is a rod-like structure probably consisting of two alpha helices coiled around each other like a two-stranded rope. In intact muscles, LMM and HMM perform entirely different functions.

Thick filaments are formed by packing of myosin molecules. About 400 molecules bind one another and their double heads project at interval distances of 14.3 nm along the fibre axis, staggered at 120° intervals so that every fourth pair of heads is aligned with each other at 43 nm. Myosin filaments are oriented in each direction from the M line lying in the middle of the H zone. A special protein *(M-protein)* at the M line marks the midpoint at which the myosin fibres meet tail to tail.

(b) *Actin:* Actin is a water-soluble, globular protein molecule consisting of 376 amino acid residues, including one residue of N^3-methylhistidine. As isolated, there is one molecule of ATP bound per actin.

Actin also occurs in two forms. In the monomeric form the actin molecules are called globular actin molecules, or *G-actin*. They lack the ATPase activity, but each molecule possesses a molecule of ATP firmly bound to it. Each molecule of G-actin has a formula weight of about 45,000. In the polymeric form the actin molecules are fibre-like and are called fibrous actin, or F- *actin*. They occur when globular actin molecules undergo polymerisation in solutions of high ionic concentrations. The F-actin molecules are in the form of a two-stranded rope. The G-actin molecules can stick together only if the ATP bound to them becomes hydrolysed to ADP with the release of energy.

$$\text{G-actin-ATP} \longrightarrow \text{F-actin-ADP} + \text{Pi}$$

In mammalian skeletal muscle tissue *myosin and F-actin* are arranged to form myofibrils. Along the main axis of the myosin unit are the projections know as the side chains. They are distributed around the myosin in a spiral manner and point towards actin units. In addition to its structural role, actin has the important property of activating the myosin ATPase, which catalyses ATP hydrolysis. This function of actin is important in the hydrolysis of ATP by myosin that accompanies muscle contraction.

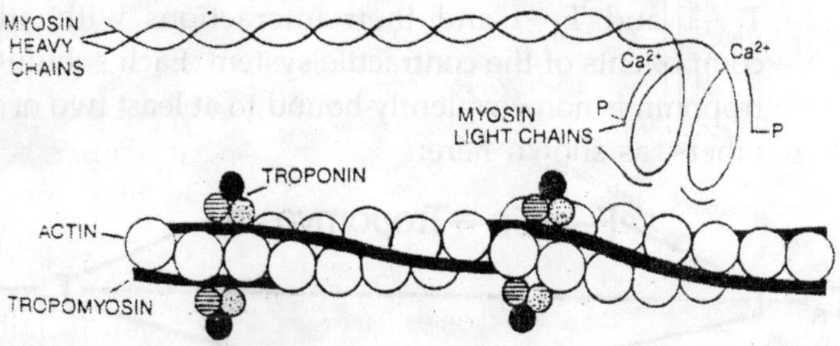

Diagrammatic representation of relative positions of Tropomyosin, troponin, and actin in thin filament of muscle. Each troponin comprises 3 different subunits. Tropomyosin lies in the groove of the helix, which is undisturbed. There is one tropomyosin and one troponin for each seven pairs of G-actin (after Adelstein and Eisenberg 1980).

(c) *Tropomyosin:* Tropomyosin is a rod-shaped molecule (MW = 66,000) about 41 nm long and consists of two non-identical, α-helical polypeptide chains (α and β) twisted around each other in a coiled coil. The rod shaped molecules lie end to end in the grooves made by the two strands of actin filaments. This is present in lesser amount than actin and myosin. It lacks the ATPase activity. As said before, tropomyosin plays an important role in sensitising the contractile proteins to calcium. Tropomyosin A (sometimes called Paramyosin) and Tropomyosin B are the two known forms of tropomyosin. Tropomyosin is found as a contaminant of highly purified actomyosin preparations and is associated with another protein, troponin that has been shown to combine with actin and exert through calcium an important effect on the reactions of the contractile system. This sensitivity is not shown by pure actin and myosin.

(d) Troponin: Troponin molecule (MW = 76,000) contains three different subunits, each named according to its function. T_n-T subunit binds tropomyosin, T_n-I inhibits the interaction between myosin and actin, and T_n-C subunit binds Ca^{2+} and regulates the interactions among T_n-T and T_n-I and their interactions with other components of the contractile system. Each subunit of troponin is non-covalently bound to at least two of the others, as shown here:

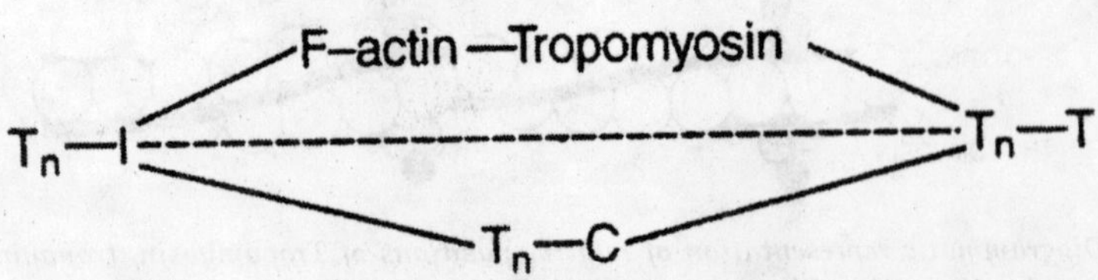

T_n-T (MW = 37,000) contains 259 amino acid residues in nown sequence, half of which are either positively or egatively charged. The latter are clustered in the NH_2 terminus, the former in the COOH terminus. T_n-T binds strongly to tropomyosin as well as to T_n-C, and serves to anchor T_n-I and T_n-C to the actin-tropomyosin complex.

T_n-I (MW = 20,000) contains 179 amino acid residues, including a region rich in basic amino acids that contains the actin-binding site. Its NH_2- terminal contains the T_n-C-binding site. In resting muscle, T_n-I inhibits the interaction of myosin and actin and prevents actin activation of myosin ATPase activity.

T_n-C (MW = 18,000) contains 159 amino acid residues rich in glutamic and aspartic acids. Its sequence is highly homologous to that of calmodulin, and like which it contains four divalent ion-binding sites. Two of the divalent ion-binding sites of T_n-C are specific for Ca^{2+}, but Ca^{+} and Mg^{2+} compete for binding at the other two. Under physiological conditions,

T_n-C is thought to bind two Ca^{2+} and two Mg^{2+} ions. Indeed, T_n-C and calmodulin appear to participate equally well in glycogenolysis in skeletal muscle.

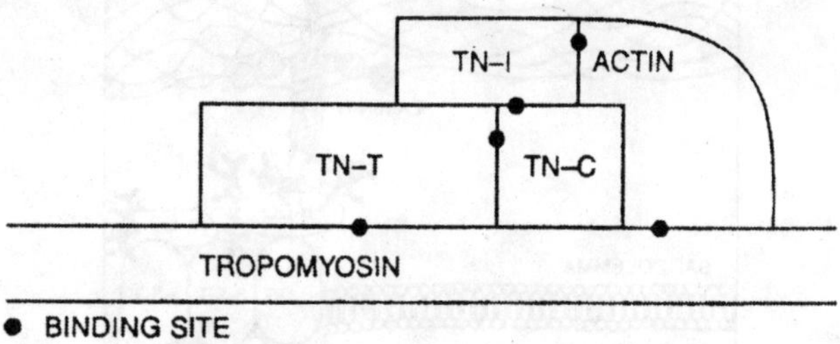

Diagrammatic relationship between actin, tropomyosin and three subunits of troponin molecule.

T_n-T, T_n-I, but not T_n-C, are covalently phosphorylated by cAMP-dependent protein kinase and ATP. The physiological significance of phosphorylation of T_n-I and T_n-T in skeletal muscle is not yet explained. However, in cardiac muscle, phosphorylation of T_n-I is accompanied by a decrease in actin activation of myosin ATPase activity. This may explain the increase in the rate of relaxation of cardiac muscle after administration of epinephrine.

Dystrophin-Glycoprotein Complex and it Application

A large protein called dystrophin (MW 427,000) forms a rod that connects actin to two other proteins in the sarcolemma and they in turn connect to the merosin subunit of the adhesion molecule laminin in the extracellular matrix. These two proteins in the sarcolemma are α -DG (dystroglycan) and β -DG forming the dystroglycan complex. This complex is in turn associated with the sarcoglycan complex consisting of three glycoproteins, a, P, and y sarcoglycan.

It appears that the dystrophin-glycoprotein complex provides structural support and strength to muscle fibril, but it may have some additional function also. The complex is of special interest because congenital defects in it cause many of the different forms of muscular dystrophy (Worton, 1995). The dystrophin gene, which is one of the largest known genes, is also expressed in smooth and cardiac muscle.

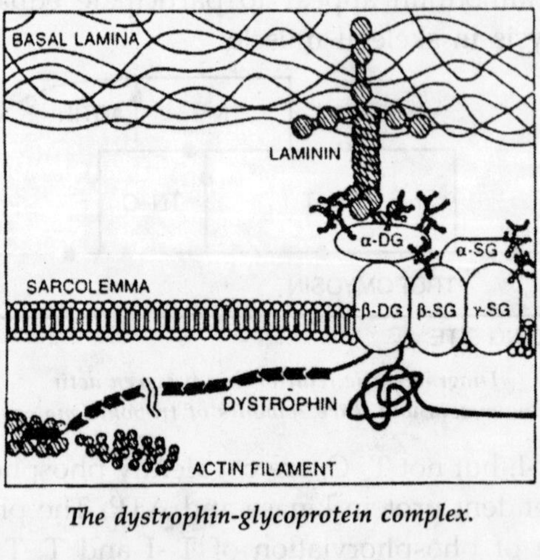

The dystrophin-glycoprotein complex.

Mutations in the gene coding for the various components of the dystrophin-glycoprotein complex cause muscular dystrophy, a syndrome characterised by progressive muscle weakness. Some forms of this disease cause great disability and are eventually fatal. In Duchenne's muscular dystrophy, which is x- linked, dystrophin is absent from the muscles. It occurs in about 1 of every 3000 male infants and is usually fatal by the age of 30. In Becker's muscular dystrophy, dystrophin is present but altered or reduced in amount. Autosomal recessive muscular dystrophy is due to a mutation in the gene for Y - sarcoglycan. In some cases mutations in one or another of the genes for all the three forms of sarcoglycans are associated with late-onset limb-girdle muscular dystrophy. However, the ways in which the various abnormalities of the dystrophin-glycoprotein complex cause dystrophy are at present unknown.

Role of Cross-Bridges

The actin (thin) and myosin (thick) filaments are linked together by cross-bridges which play an important role in muscle activity. When a longitudinal section of muscle is examined under the electron microscope, it can be seen that the cross bridges project outward from both the ends of a myosin filament at fairly regular interval of 60 or 70Å. In the centre of the myosin filament there is a bridge-free region, about 0.15 to 0.2 μ long. This absence of cross-bridges is responsible for the pseudo H zone in the centre of A bands.

It has been recently shown by *H.E. Huxley* (1965) that the cross-bridges are nothing except the heads of the myosin molecules and all such heads (*zigzag* lines) or the cross-bridges have the enzymatic and actin-binding properties. During muscle activity each cross-bridge is attached to an active site on the actin filament at a definite rate and then detached from the same. Thus, cross-bridges play an important role in muscle contraction.

Physiology of Muscle Contraction: Muscles contract in response to a stimulus that is delivered to them through a motor nerve. So the first step in the excitation of a muscle is the arrival of a nerve impulse along the nerve fibre that is connected to the muscle at a junction called the *"end plate"*. In the resting state the outer membrane of the muscle fibre is electrically polarised, *i.e.*, its outer surface is positively charged with respect to the inner surface. The arrival of the nerve impulse brings about the release of *acetylcholine* at the neuromuscular junction. This substance depolarises the muscle membrane by increasing its permeability to Na^+ ions and this depolarisation in turn excites an *"action potential"* which propagates along the muscle fibre. The action potential sets off two independent electric currents in the muscle fibre, of which one is a weak longitudinal current and which, according to *Ove Knudsen* (1954) does not induce contraction. But the other current is a transverse wave that moves inward along the system of transverse tubules and is said to be responsible for contraction.

The following theories or models have been put forward to explain the mechanism of muscle contraction.

Theory of Sliding-Filament

According to sliding-filament or Interdigitation theory, proposed independently by *A.F. Huxley* and *R. Niedergerke*, and by *J. Hanson* and *H.E. Huxley* (1958), the two kinds of filaments slide past each other so as to produce change in the length of the voluntary muscle. The theory proposes that shortening of the muscle fibre is caused by a *sliding movement of interdigitating filaments*, bringing about an increase in the overlap between them.

It has been found that during both contraction and stretching, the length of A-bands remains constant and the length of the I-bands changes in accordance with the length of the muscle. Since the length of the A-band is equal to the length of the thick filaments, so the length of these filaments, also remains constant. But the length of H-zone increases or decrease with the length of the I-band, so that the distance from the end of one H-zone through the Z-line to the beginning of the next H-zone remains almost the same. This distance is equal to the length of thin filaments, so they too do not change their length to a large amount. The only conclusion which has been drawn from these observations is that, when the muscle changes its length, the filaments themselves do not perceptibly change in length but the two sets of filaments slide past one another, *the thin filaments moving farther into the A-bands during contraction and moving farther out of them during stretching.* So, in a powerful contraction the ends of thin filaments first meet each other and then they crumple up or overlap. Later on the same may happen with the thick filaments and the crumpling produces new band patterns in the muscle.

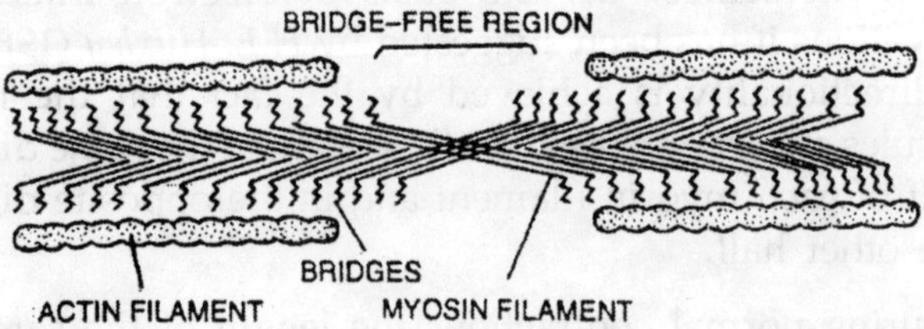

Schematic representation of the contact of actin and myosin in a voluntary muscle. The thin actin filaments above and below are so shaped that certain sites are closest to thick myosin filament in the middle. The heads of individual myosin molecules (zigzag lines) extend as cross bridges to the actin filament at these close sites (after H.E. Huxley, 1965).

The cross-bridges play an important role in muscle contraction and in fact the sliding movement is mediated by the bridges. In fact, mechanical linkage between the two types of filaments occurs via cross bridges. It has been concluded by *H.E. Huxley* (1965) that the heads of myosin molecules, which have the enzymatic and actin-binding properties, serve as the cross-bridges connecting the actin and myosin filaments in the muscle. Besides, the myosin molecules possess an inherent directionality because they aggregate with their heads pointed in one direction along half of the filament and in the opposite direction along the other half of the myosin filament.

This property of myosin molecules is very important because it explains a crucial feature of the sliding-filament theory at the molecular level. In a sliding-filament system in which a relative force is developed between actin and myosin molecules of the two types of filaments, it is essential that the appropriate directionality of sliding be built in some way.

During contraction the thin filaments move towards each other in the centre of the A bands, so that it is required that all the elements of force generated by the cross-bridges in one half of the A band be oriented in the same direction, and that the direction of the force be reversed in the other half.

Thus, during interaction of actin and myosin the direction of the force developed depends on the orientation of actin and myosin molecules. On the basis of electron microscope observations it has been suggested by *H.E. Huxley* (1965) that this directionality is achieved by the fact that the myosin molecules are arranged so that they point in the same direction in half of each myosin filament and in the opposite direction in the other half.

During normal contractions the length of the sarcomere changes about 20 per cent and thin filaments in each half of the A band have to move about a quarter of a micron. Hence it seems physically impossible that the cross-bridges could remain attached to the same point on the actin filament throughout this process. It has been explained by *H.E.Huxley* (1965) that the bridges are attached to one site on the actin filament for part of the contraction, then detach and reattach themselves at a new site farther along. Moreover, at each site where cross-bridge and actin filament interact, one molecule of ATP is split to generate a sliding force between the two kinds of filaments.

The interaction of filaments during sliding has been explained by giving the analogy of a ratchet, and the way in which the cross-bridges of myosin hook on to consecutive active sites on the actin filament further supports this analogy. As the skeletal muscle can contract by some 30 per cent, so each individual cross-bridge must detach itself from one site on the actin and reattach itself to another site further along, repeating the process 5 or 6 times, "with an action similar to a man pulling in a rope hand over hand" (*Wilkie,* 1968). However, another concept is that electric fields are established by the cross-bridges which attract the actin filaments thereby causing them to move.

Davie's Principle

According to *R.E. Davies* (1963), the side chains on the myosin filament are important in this theory. Each of these side chains terminates in a protein-bound molecule of ATP. Near the base of the side chain is a fixed negative charge probably –COO⁻ of the side group of glutamic or aspartic acid. Repulsion between this fixed charge and the net negative charge at the tip of ATP is thought to keep the side chain extended in the resting molecule. Otherwise the side chain polypeptide would tend to coil up. Should it do this, according to the theory, the phosphoric anhydride system would be drawn into a region of the myosin that actually provides a local catalyst for hydrolysing such anhydrides.

The side chain on myosin is called a cross-bridge because it can be part of an ionic bond or bridge to a neighbouring actin molecule. This bridge is completed in the moment of activation when C^{2+} ions are released from a region of the cell called the sarcoplasmic reticulum. This ion with two plus charges can fit between the two sites with minus charges, one on myosin and one on actin. Once this happens, the minus charge at the ATP tip is neutralised. The force of repulsion between this tip and the fixed negative charge at the base of the side chain is removed.

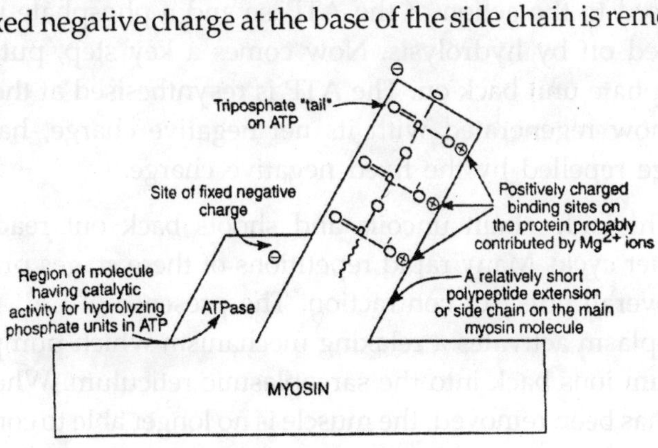

The cross-bridge in muscle myosin is depicted here in its resting or extended state. The net charge at the ATP tip is-1. Repulsion between it and the fixed charge keeps the side chain extended. The side chain is called H-meromyosin.

The side chain at once coils into a helix. It cannot do this, of course, without pulling on the actin molecule, for as long as the calcium ion is there the actin and myosin are for all practical purposes bound together indicates that this "pull" can only cause the actin and myosin filaments to slide by other, as *Huxley* said they must.

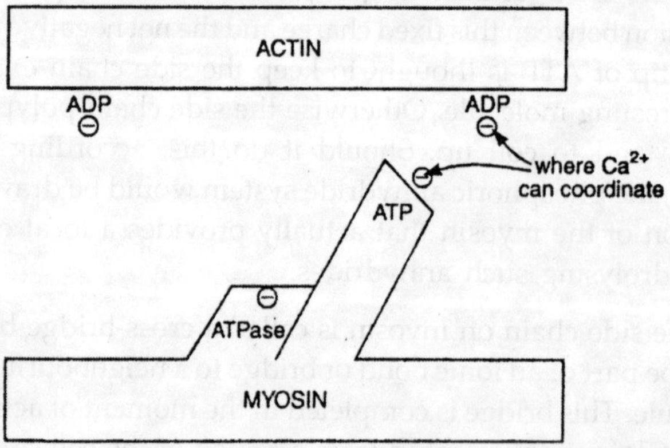

Showing the arrangement of actin and myosin in the resting state. When the muscle is activated, calcium ions are released from the sarcoplasmic reticulum, and they move to the negatively charged sites as shown.

When the side chain is in its coiled phase, the ATP tip is exposed to the action of the ATPase and a phosphate unit is clipped off by hydrolysis. Now comes a key step, putting a phosphate unit back on. The ATP is resynthesised at the ATP tip, now regenerated with its net negative charge, has this charge repelled by the fixed negative charge.

This side chain uncoils and shoots back out ready for another cycle. Many rapid repetitions of these stages produce the overall effect of contraction. The presence of Ca^{2+} in the sarcoplasm activates a relaxing mechanism which pumps the calcium ions back into the sarcoplasmic reticulum. When the Ca^{2+} has been removed, the muscle is no longer able to contract.

The *Huxley* model of muscle contraction is thus given a molecular interpretation by the theory of *Davies*.

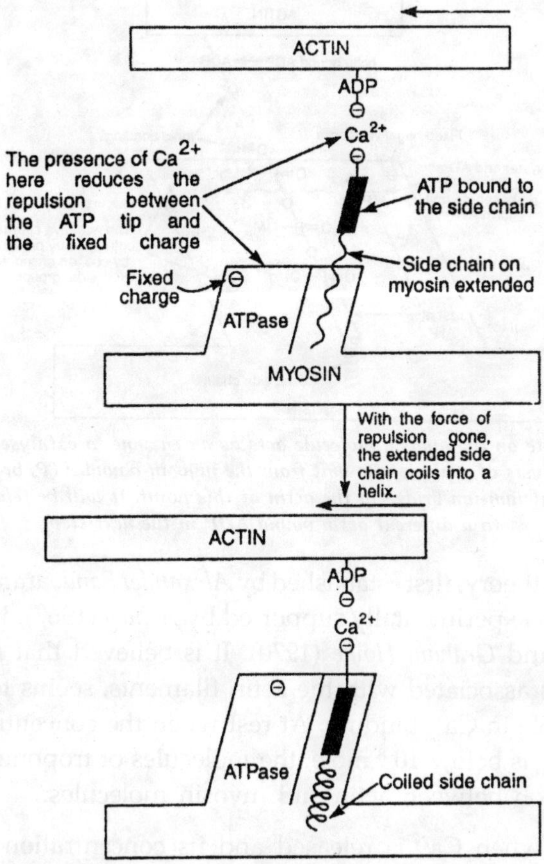

The calcium ion removes the force that keeps the side chain extended, and it coils back towards the main chain of myosin. As it does so, the actin filament is pulled in the direction of the arrows, causing it to slip along the myosin chain.

Principle about Calcium Release

The theory holds that contraction is initiated by Ca^{2+} ions released from a binding site in the muscle (sacs of the sarcoplasmic reticulum) by the action of electrical stimulus and relaxation results by subsequent uptake of these released Ca^{2+} ions by the sarcoplasmic reticulum. Thus, Ca^{2+} is considered to activate the contractile machinery.

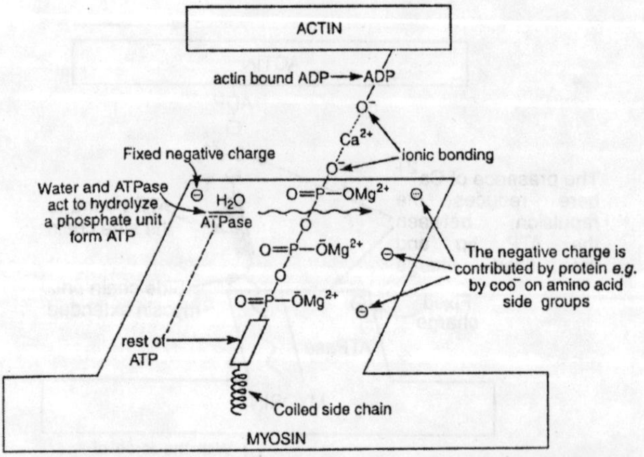

A site on the myosin molecule acts as an enzyme to catalyse the hydrolysis of a phosphate unit from the myosin-bound ATP, breaking the calcium-ion bridge to the actin at this point. It will be reformed to a different actin bound ADP in the next step.

This theory, first established by *Alexander Sandow* and others has been experimentally supported by *Ashley* (1967), *Winegrad* (1968), and *Graham Hoyle* (1970). It is believed that *troponin,* which is associated with the actin filaments, seems to play a major role in Ca^{2+} binding. At rest when the concentration of free Ca^{2+} is below 10^{-8} mole, the molecules of troponin inhibit interaction between actin and myosin molecules.

But when Ca^{2+} is released and its concentration rises, it combines with the troponin. Thereupon the restraint that troponin places on the interaction of actin and myosin is released and the two types of filaments slide past one another causing the muscle to contract. Then the removal of Ca^{2+} during relaxation reverses the process.

However, there are two sources of Ca^{2+}. One is the *internal calcium* that supplements the other source, the *external calcium* entering the fibre during excitation. The possible sources of internal Ca^{2+} include the membranes of the fibre surface, the clefts, the invaginated tubules, the sarcoplasmic reticulum, and the mitochondria. The sarcoplasmic reticulum sacs adjoining the invaginated tubular system are considered as the major source of supplementary internal Ca^{2+}.

The release of the internal Ca^{2+} is caused by the initial stimulus which depolarises the fibre membrane and the conduction of the negative depolarising charge down the transverse tubules. It has been shown by *Hygo-Gonzalez-Serratos* that the rate of inward conduction along the tubules of a frog fibre is about 8 cm/sec. The ions pass down the tubules and across the junction between the tubules and the sarcoplasmic rericulum.

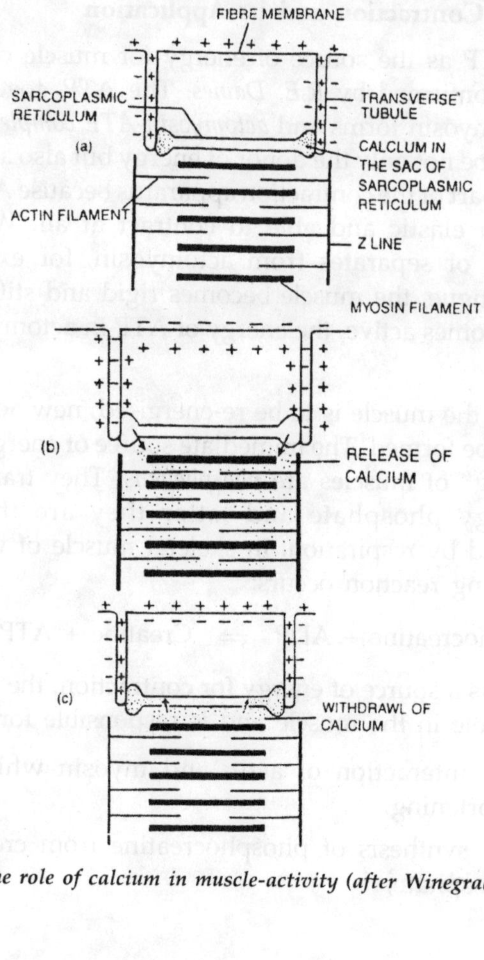

The role of calcium in muscle-activity (after Winegrad).

The process of relaxation depends on the return of the charge on surface membrane to its original condition. As the excitatory depolarisation is followed by a rapid repolarisation, the stimulus to the release of Ca^{2+} is quickly turned off and hence the release of Ca^{2+} is stopped. In other words, the relaxation is due to the diminishing amount of free Ca^{2+} as it is rebound to the reticulum. However, by the time the contraction reaches its peak, the free Ca^{2+} has already fallen to a low level. Thus muscle contraction is turned on by the release of Ca^{2+} and turned off by its withdrawal.

Energy of Contraction and its Application

The ATP as the source of energy for muscle contraction was first confirmed by *R.E. Davies*. The ATP, together with actin and myosin forms and *actomyosin-ATP complex*. Thus, it appears to be not only the donor of energy but also a necessary structural part of the contraction apparatus because ATP makes actomyosin elastic and able to contract at all. When ATP disappears or separates from actomyosin, for example, in extreme fatigue, the muscle becomes rigid and stiff. When a muscle becomes active, the energy of ATP in actomyosin-ATP is used up.

If then the muscle is to be re-energised, new actomyosin-ATP must be formed. The immediate source of energy for such "recharging" of muscles are *phosphagens*. They transfer their high-energy phosphate and latter they are themselves re-energised by respiration. In skeletal muscle of vertebrates the following reaction occurs:

$$\text{Phosphocreatine} + \text{ADP} \rightleftharpoons \text{Creatine} + \text{ATP}$$

Thus, as a source of energy for contraction, the ATP plays a central role in the muscle and is responsible for:

(i) the interaction of actin and myosin which causes shortening,

(ii) the synthesis of phosphocreatine from creatine and phosphate,

(iii) the synthesis of glycogen from lactic acid. This is necessary because the conversion of glycogen to lactic acid is another source of energy to synthesise ATP other than the phosphocreatine reaction.

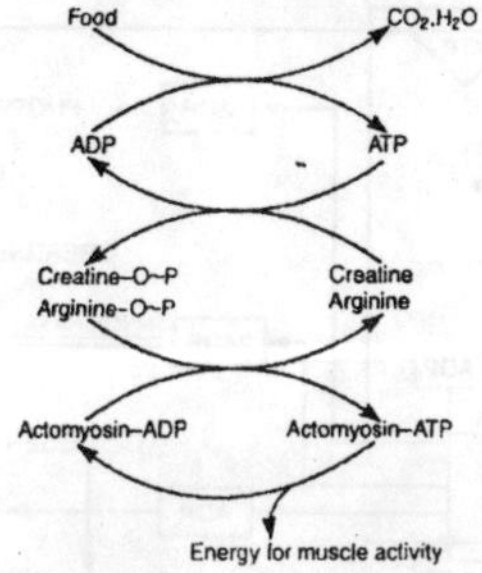

Showing that respiration supplies energy for muscle-activity via the phosphagens.

Chemical Reactions in Muscle Contraction: The various chemical reactions and the flow of energy in the muscle during contraction have been described by *Loewy* and *Siekevitz* (1966). Firstly myosin reacts with ATP to form a myosin-ATP complex. Then in the presence of K^+ and F-action, actomyosin-ATP complex is formed.

This complex in the presence of Ca^{++} and ATPase splits the ATP to inorganic phosphate and actomyosin-ADP complex, which then breaks down to free actin, free myosin and ADP. Thus, now it is known that the muscle contraction is turned on by the release of Ca^{++} and the hydrolysis of the ATP of actomyosin-ATP complex provides the necessary energy for the two kinds of filaments to slide past each other during contraction.

The relaxation is caused by binding Ca^{++} back into the sarcoplasmic reticular sacs, and thus inhibiting the breakdown of actomyosin-ATP complex. It has been also proposed that the Ca^{++}, as a trigger for muscle contraction, performs its work by combining with troponin which otherwise puts a restraint on the interaction of actin and myosin.

The muscle activity is continued by the continuous removal

of ADP that accumulates in contraction, or in other words, by the conversion of ADP to ATP which is done in several ways as shown in Fig.

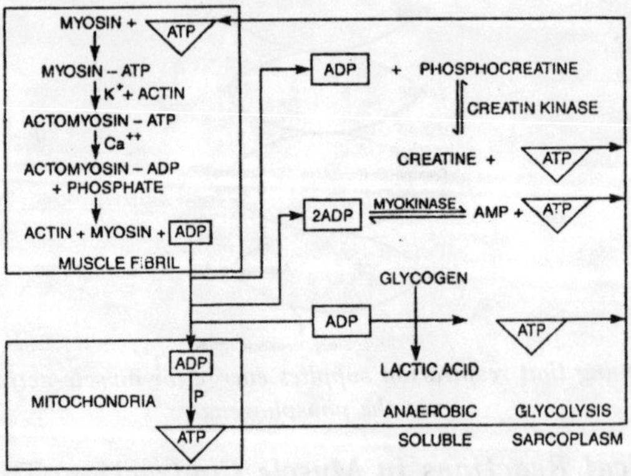

The chemical reactions in muscle contraction and the flow of energy.

Summary of the Present Concept: According to our present information the important events leading to muscle contraction are as follows.

(a) The arrival of an impulse
(b) Reduction of electric potential or depolarisation of the fibre membrane
(c) Release of calcium (the activating substance)
(d) ATPase activation
(e) Hydrolysis of ATP at each site where cross-bridge and actin filament interact to generate a sliding force between the two kinds of filaments.

Similarly, the processes associated with relaxation are:

(a) Exhaustion of the nerve impulse
(b) Withdrawal of calcium
(c) Recreation of electric potential or repolarisation of the fibre membrane.
(d) ATPase inhibition.
(e) Resynthesis of ATP.

Nerve impulse & Release of Ca^{++} Reduction of electric potential ATPase activation

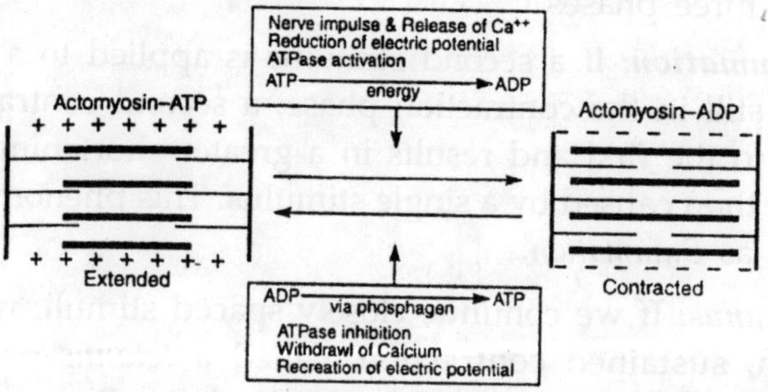

The summary of events leading to muscular contraction and relaxation.

The Single Muscle Twitch: A muscle twitch can be defined as a simple single muscular contraction caused by a single short stimulus. It can be recorded by *Kymograph* and consists of three successive phases. *(i) Latent period:* As its name implies. It is the lag period or time interval between the application of the stimulus and the actual beginning of visible muscle contraction. In the frog, the latent period lasts about 0.01 second and in the figure it is indicated by the distance from A to B.

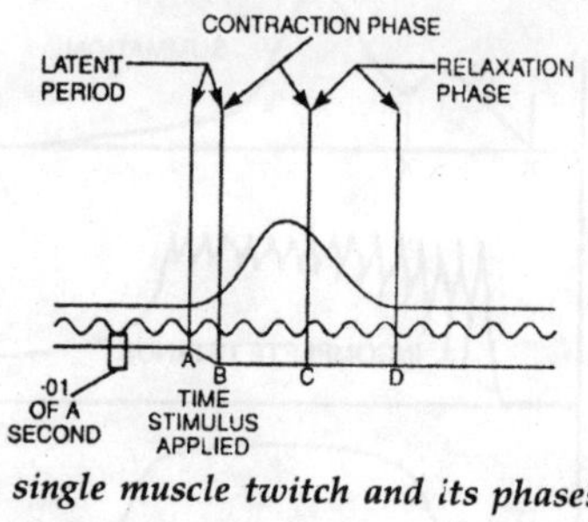

A single muscle twitch and its phases.

(ii) *Contraction Phase:* It follows the latent period, as indicated by the portion of the curve from B to C. It is the time interval in which the muscle reaches the peak of contraction and performs work. It has a duration of about 0.04 second.

(iii) *Relaxation Phase:* It is the final phase of the muscle twitch, or the time interval in which the muscle returns to its original length. This phase lasts about

0.05 second and is represented by the portion of the curve between C and D. Thus, it is the longest of the three phases.

Summation: If a second stimulus is applied to a muscle that is still in the contraction phase, a second contraction is added to the first and results in a greater shortening of the muscle than caused by a single stimulus. This phenomenon is known as *Summation.*

Tetanus: If we continue closely spaced stimuli, we get a smooth, sustained contraction, called a *tetanus* or tetanic contraction. A complete tetanus may be defined as a sustained contraction of a muscle due to the fusion of many twitches following each other in rapid succession.

In case the successive stimuli arrive at the muscle with long time intervals between them, the individual contractions can still be recognised since some degree of relaxation occurs between them to give a curve with a wavy plateau as shown in Fig. This condition is called *incomplete tetanus.*

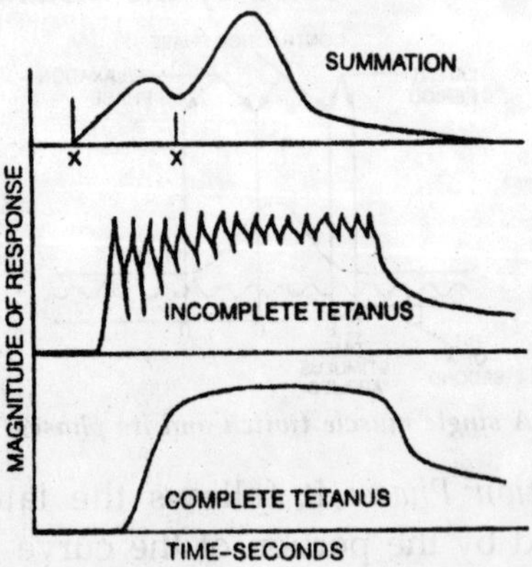

Diagram of summation, incomplete tetanus and tetanus.

Definition of Refractory Period

Skeletal muscle, like nerves, exhibits a refractory period. It is an extremely short time interval lasting about 0.002 to

0.005 second, during which a muscle will not respond to a second stimulus when given immediately after the first one. The second contraction is superimposed upon the first to give the summation effect.

It is thought to represent the time necessary for the occurrence of various physical and chemical changes which will allow subsequent contraction. Cardiac muscles have a longer refractory period than that of skeletal muscle and, therefore, lasts during the entire period of systole (the contraction of atria or ventricles of a heart) which means that cardiac muscle will not respond to a second stimulus while it is contracting.

Tiredness of Muscle

If the energy is not continuously supplied to a contracting muscle, the actomyosin-ATP complex cannot be regenerated and the muscle becomes fatigued. Thus, fatigue is a condition or state whereby a muscle experiences diminished ability to contract as a result of prolonged stimulation. However, it rarely occurs in intact muscle of the body, but can be easily demonstrated in excised muscles. Several factors are responsible for muscle fatigue, *e.g.*, lack of oxygen, lactic acid accumulation, and decreased level of phosphocreatine.

Significance of Rigor Mortis

When muscle fibres are completely depleted of ATP and phosphorylcreatine, they develop a state of rigidity called *rigor*. When this occurs after death, the condition is called *rigor mortis.* As we know soon after death the body passes into rigor mortis or the so-called stiffening of death. Normally, when muscle passes into rigor mortis its ATP and the glycogen content disappear and lactic acid accumulates in it. However, the mechanism by which muscle contracts after death (or in the absence of the action potential) is not known.

Types of Contraction

1. *Isotonic Contraction:* Isotonic means "same tension". It is that contraction which occurs against a constant load and there is a change in the length but the tension of the muscle remains practically the same.

The velocity of shortening as a function of load (force velocity relation) is given by the equation:

$$V = (P_o - P)\, b/(P + a)$$

where V is velocity of shortening, P_o is the maximum tension developed by the muscle, P is a given load, a is a constant with dimensions of a force, and b is a constant with dimensions of velocity. According to *Hill* (1953), the velocity of shortening depends on the difference between the force on the muscle and the maximal force it can develop. Thus, mechanical work is done in isotonic contraction.

The overall mechanical efficiency of skeletal muscle (workdone/total energy expenditure) ranges upto 50 per cent while lifting a weight during isotonic contraction and is essentially 0 per cent during isometric contraction.

2. *Isometric Contraction:* Isometric means "same measure" or length. When a muscle can contract without changing its length, it is called isometric contraction. In this case the tension increases sharply, but no mechanical work is done.
3. *Heat production in muscle:* In view of the laws of thermodynamics the energy supplied to a muscle must equal its energy output. The energy output appears in work done by the muscle, the energy-rich phosphate bonds formed for subsequent use and in heat. The contraction of a muscle is accompanied by the liberation of heat in two phases:
   (i) The *initial heat* is liberated during the contraction-relaxation cycle and is divided into several stages

like *activation heat* (maintenance heat in Sustained contractions), *shortening heat* (liberated during the actual shortening of the muscle and is proportionate in amount to the distance the muscle shortens), and the *relaxation heat* (produced at the end of an isotonic contraction when a muscle has done work by lifting a load).

(ii) The *recovery heat* or *delayed heat* is liberated in the period immediately following the contraction cycle. This recovery heat is the heat liberated by the metabolic processes that restore the muscle to its pre-contraction state. The recovery heat is approximately equal to the initial heat, *i.e.*, the heat produced during recovery is equal to the heat produced during contraction.

The difference between the heat production of an isotonic and an isometric contraction is the shortening heat. The fact that shortening heat is independent of the load indicates that when a muscle performs more work, energy must be produced by the muscle. This is the *extra energy liberation,* and the phenomenon is often called the *Fenn effect*.

Physiology of Amoeboid and Ciliary Movements: Microbes and small multicellular organisms have other modes of locomotion, of which *amoeboid* and *ciliary* movements are the most important.

Amoeboid Motion

The characteristic movement of *Amoeba* by pseudopodia is called amoeboid movement. It is also found in the white blood corpuscles and in amoebocytes of sponges.

When a pseudopodium is formed, a blunt, finger-like projection called the *hyaline cap,* composed only of ectoplasm, first appears. A little later the *plasmasol* flows into this projection as it extends forward. At the start of the movement, *Amoeba* usually forms several small pseudopodia, of which one is gradually enlarged while the others disappear.

The protoplasm which enters the pseudopodium is naturally withdrawn from other parts, so that the animal not only changes its body shape but also its position. Thus, pseudopodia bring about a change in the shape and position of the *Amoeba.*

Theories of Amoeboid Movement

1. *Contraction Theory:* According to *Jennings* (1904), who worked on *Amoeba verrucosa,* a rolling movement of the plasmalemma accompanied by protoplasmic streaming, brings about locomotion. A similar contraction theory has been put by *Dellinger,* who worked on *Amoeba proteus,* suggesting a "walking" movement. In other species, *e.g., Difflugia,* there may be a walking movement, wherein the advancing tip is attached and the remainder is pulled forward.
2. *Change of Viscosity Theory:* This is also called *Gel-Sol-theory* of *Mast* and *Pantin.* According to this theory the two states of cyctoplasm are continuously converted into one another and bring about the formation of pseudopodia. Since the change of the gel form into the sol occurs continuously, a current of plasmasol is maintained towards the anterior direction.

   This flow itself seems to be brought about by the contraction of the plasmagel tube at the posterior end, thus forcing the plasmasol towards the anterior tip. During locomotion the elastic and contractile force of plasmagel is maximum at the sides, intermediate at the posterior end, and minimum at the anterior end. This results in an elongated shape of the animal and a forward extension of the anterior end to bring about locomotion.

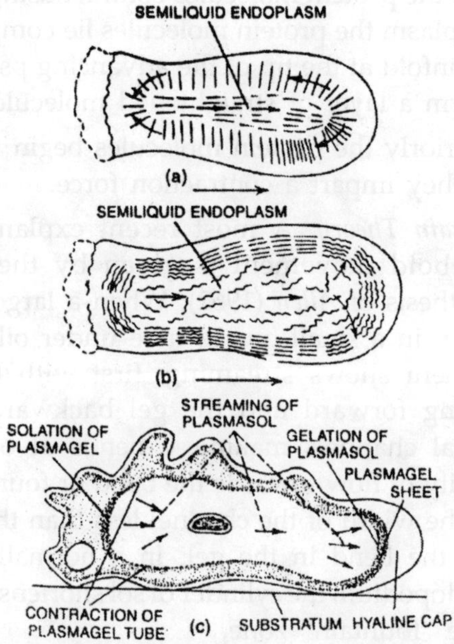

Theories of Amoeboid movement, (a) According to Allen's theory, the Amoeba is pulled forward by the contraction of cortical gel near the head, (b) According to Goldacre, contraction at the tail accounts for the movement, and (c) movement of Amoeba according to gel-sol theory of Mast and Pantin.

3. *Syneresis Theory:* According to *Pantin,* water is removed by some sort of "syneresis" from protein at the posterior end and that the forward tide of syneretic fluid is important in locomotion. The tail of some amoeboid cells *(A. Umax),* wrinkles during locomotion as if water was removed. Recent measurements of the refractive index of *A. proteus* and *A. dubia* show that in sustained locomotion the refractive index rises towards the tail, this region having 6 to 40 per cent more (protein) than the advancing tip, and this gradient of refractive index disappears on stop of movement.
4. *Theory of Folding and Unfolding of Protein Molecules:* According to *Goldacre* (1951), the amoeboid movement is brought about by folding and unfolding of protein molecules and the energy for movement is supplied

when the protein molecules contract using ATP. In the endoplasm the protein molecules lie compactly folded and unfold at the tip of the advancing pseudopodium to form a layer of straightened molecules.

Posteriorly the protein molecules begin to fold again and they impart a contraction force.

5. *Fountain Theory:* A most recent explanation of the amoeboid movement is given by the "fountain" hypothesis of *Allen* (1962). When a large *Pelomyxa* is broken in a small quartz tube under oil the isolated fragment shows streaming, first with the axial sol flowing forward and the gel backward/then with several channels making U-bends at one end. The velocity of flow towards the bend or fountain is faster and the width of the channel less than the flow away from the bend in the gel. In a normally advancing pseudopodium the cylinder of sol shortens and thickens in the "fountain" zone.

It is postulated that gel strands, attached to the rim of the advancing tube, contract and pull the anterior sol forward. Water is removed to the hyaline cap by syneresis and passes back under the plasmalemma to be returned to the protein at the posterior zone of recruitment of sol.

Flow is chiefly at the boundary or shear zone between sol and gel. It is argued that contraction at a U-bend between sol and gel can equally well explain amoeboid movement and cyclosis.

Ciliary and Flagellar Movements: Cilia are the motor organelles differentiated at the cell surface. They are often classified into two types.

(i) *flagella*, which are relatively larger organelles occurring usually singly or in small numbers upon cells. Typical flagella characterise the class mastigophora (Gr. *mastix*-whip, *pherein* = to bear) of the phylum Protozoa.

(ii) *cilia* proper, which are relatively much smaller and

occur characteristically in large numbers upon each cell. Cilia proper are seen in ciliated protozoans, coelenterates, and turbellarians.

Ciliary activity is restricted to an aqueous medium, and hence is found only on the surfaces which are submerged, or atleast covered by an aqueous film. In some cases cilia are concerned in one or more numerous functions, such as feeding (rotifers, protozoans, protochordates, etc.), circulation of body fluids (annelids), cleansing (sea anemones), respiration (molluscs, protochordates, etc.) and the movements of materials within ducts and tubules (coelenterates, echinoderms, and molluscs.)

Physiology of Cilia

Cilia of living cells represent a great uniformity in structure. Each cilium consists of a sheath containing an axial filament that usually extends distally beyond the limits of the sheath. The sheath may be circular or ovoid in cross-section and the axial filament pursues a straight or spiral course through the sheath. The proximal end of the axial filament is always provided with a basal granule, which is derived from the cell centrosome. Many cytologists have described a system of fibrils running between the basal granule and cell nucleus.

The ultrastructure of a cilium has been described with the aid of electron micrographs by a number of workers including *Roth* (1956) and *Porter* (1957). A cilium has nine parallel fibrils surrounding a pair of axial fibrils. The "9 + 2" arrangement of microtubules seen in cross section of a cilium (or flagellum) is one of the most conserved organelle structures in eukaryotic cells. The outer fibrils, with a diameter of 250 to 500 Å, are about twice as thick as the axial ones. All these observations, together with the fact that cilia show positive form and intrinsic birefringence, indicate that longitudinally oriented submicroscopic micelles compose the cilium. In this characteristic, therefore, cilia resemble with the contractile fibrils of muscle tissue.

Significance of Ciliary Control and Coordination

Cilia possess a well-developed *automaticity.* This automaticity of ciliary action resides entirely in the cilium and an immediately adjacent region of the cell. An organic connection with the basal granule is essential to the beat. On breaking the connection with the basal granule is essential to the beat. On breaking the connection between the cilia and granules, the ciliary activity is immediately stopped. It is not yet fully established whether the basal granules themselves, or some other functional elements in the same general area, are the essential elements for normal rhythmic activity.

Metachronism, of the ciliated surface is of great importance. In metachronal rhythm each preceding cilium beats a little ahead to its succeeding cilium. In *Opalina,* the direction of the metachronal wave changes freely with the direction of swimming, passing backward, to right or to left. Ciliary activity normally involves the use of oxygen, the rate of consumption varying with the ciliary rate.

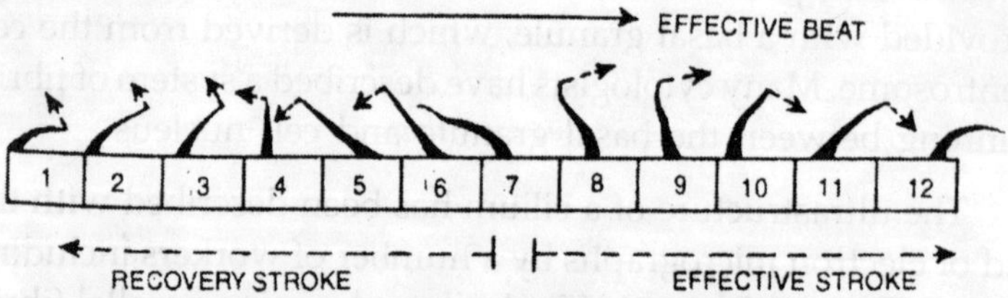

Diagrammatic representation of the metachronal rhythm. Cilia 1 to 12 are beating in sequence.

Types and Characteristics of the Ciliary Movements: According to *Gray* (1928), these are Pendular, Flexural, Undulatory and Funnelform or three-dimensional movements. Ciliary activity sometimes involves a combination of pendular and flexural activities (*e.g.*, the frontal cilia of the gill of *Mytilus*. Forward locomotion in such common flagellates as *Euglena,* in which the flagellum was formerly believed to pull the organism and hence function as a tractellum, appears to result from a bending of the whole flagellum, or at least of its active tip backward.

The undulatory activity typically passes from the base to tip around the flagellum, as well as along it. Such movement imparts rotational and gyrational components, along with a usual backward thrust. The resulting rotation and gyration of the body of the flagellate around an axis which constitutes the direction of locomotion is considered to provide an important force propelling the organism forward. The body proper acts under these circumstances as a screw propeller.

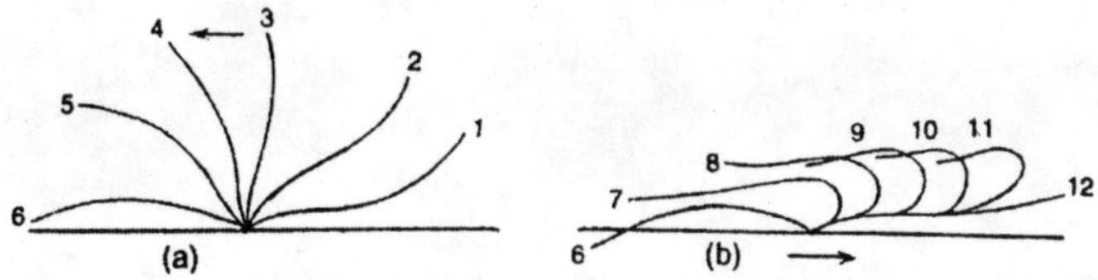

Movement of a frontal cilium of Mytilus (a) effective, and (b) recovery strok.

Factors Influencing the Ciliary Activity: Ciliary activity is influenced by a number of factors including temperature, hydrogen ion concentration, amount of oxygen, and other cations and anions in the medium.

2

System of Respiration

The term *respiration* includes both oxidative metabolism within the cell, and the means by which oxygen reaches the cell. Since all oxidative metabolism involves the production of carbon-dioxide, the transport and elimination of this gas is normally considered at the same time as the transport and utilisation of oxygen. Respiration, therefore, includes oxidating of nutrient material in the cells to release energy, and breathing.

The respiration in animals is brought about in 2 phases:

External Respiration or Breathing or Ventillation: In this phase, blood takes up O_2 from the fresh air inhaled into lungs, and gives up CO_2 to it.

Internal or Tissue Respiration: In this second phase, blood distributes O_2 to the cells and receives CO_2 from through the tissue fluid. The internal respiration is chiefly concerned with the uptake of O_2 release of energy and the production of CO_2, by the cells. It is, therefore, also referred to as the *cellular respiration.*

Respiratory System Completing Breathing

It consists of two divisions specialised for different purposes.

Respiratory Tract or Conducting Passage: Its function is piping air to and from the lungs. It comprises.

(i) Nasal cavity
(ii) Pharynx
(iii) Larynx
(iv) Trachea
(v) Two Primary Bronchi
(vi) Secondary, Tertiary Bronchi, and Terminal Bronchioles in lungs.

Respiratory Seat: The intimate interchanges between blood and air occur in this part. It consists of two lungs and their convertings, the pleura: containing within:

(i) Respiratory Bronchioles
(ii) Alveolar ducts
(iii) Atria
(iv) Alveolar sacs
(v) Alveoli

Nasal Cavity as a Part of Respiration System

The nasal cavity is the first of the respiratory organs and consists of large irregular cavity divided by a *septum*.

The *roof* is formed by the base of the skull.

The *floor* is formed by the roof of the mouth, which is composed of the maxillary bones, the palatine bones and the soft palate.

The *medial wall* is formed by the *septum* which is composed of perpendicular plate of the ethmoid bone superiorly, and the vomer-inferiorly. Anteriorly, the septum is formed by cartilage.

The *anterior wall* is formed by the nasal bones and cartilage covered with skin.

The *lateral walls* are formed by the superior maxillary bones and the inferior, middle and superior *Conchae or Turbinals* which are scroll-like (rabbit) or shelf-like projections (man), lined with respiratory epithelium. The conchae immensely increase the surface of the nasal chambers.

The *posterior wall* is formed by the posterior wall of the pharynx.

The nasal chamber are lined with a vascular, ciliated columnar epithelium, rich in mucous cells. This region of the respiratory tract acts as an air-conditioner because the incoming air is warmed, moistened, filtered and sterilised during its passage through the nose:

(1) *Warming:* Blood capillaries of the turbinals radiate heat in a similar manner as the hot-water pipes. The incoming air passing over them, therefore, gets warmed upto the body temperature.

(2) *Moistening:* The inspiratory air becomes moistened on coming in contact with the moist mucous secreted by glandular respiratory epithelium.

(3) *Cleaning:* Mucous is sticky in nature, therefore, particles of dust and other impurities in the air adhere to it. The cilia of the respiratory epithelium drive this mucous either towards the external nares or to the oropharynx where it is swallowed. Therefore, from a physiological standpoint, nasal breathing is superior to mouth-breathing.

(4) *Sterilisation:* Mucous is antiseptic in action. To some extent, it kills the bacteria of the air going to the lungs.

The *anterior* or *external nares* form the anterior opening into the nasal cavity.

The *posterior internal nares* from the posterior opening from the nasal cavity into the pharynx.

The Pharynx: The *pharynx* is a cone-shaped tubular part which is common to the respiratory system and the alimentary canal.

For descriptive purpose the pharynx is divided into three parts: (a) the naso-pharynx, (b) the oro-pharynx, (c) the laryngo -pharynx.

(a) *Naso-pharynx:* The naso-pharynx is the part which lies behind the nose above the level of the soft plate. On its lateral walls are the two openings of the *auditory lubes* which lead from the naso-pharynx to the middle ear. In its posterior wall is the lymphoid tissue known as the naso-pharyngeal tonsil or, when enlarged, as *adenoids.* These are most prominent in the children upto approximately the age of 7 years and thereafter begin to degenerate.

(b) *Oro-pharynx:* The oro-pharynx is continuous in from with the buceal cavity and below with the laryngeal part of that pharynx. The *oral tonsils* which are collections of lymphoid tissue lie in its lateral walls.

(c) *Laryngo-pharynx:* This part of the pharynx extends from the oro-pharynx above the continues as oesophagus below.

The Larynx: The *larynx* or "voice box" is situated in the mid-line of the neck between pharynx above and the trachea below. It is placed in front of the oesophagus and corresponds with the levels of the 4th, 5th and 6th cervical vertebrae. After puberty there is a considerable enlargement in the size of the larynx in the human male as compared to the female. The prominent larynx in the male is often referred to as the "Adam's apple".

The larynx consist of a framework of the following hyaline cartilages:

(a) *The Thyroid Cartilage:* This is the largest cartilage and consists of two side wings known as the laminae which are fused together anteriorly to form the *laryngeal prominence* or Adam's apple.

(b) *The Cricoid Cartilage:* This is situated below the thyroid cartilage. It is shaped like a signet ring, completely encircling the larynx. The cricoid forms the lowest part of the larynx and connects with the trachea.

(c) *The Arytenoid Cartilages:* These are two pyramid shaped cartilages situated on top of the broad part of the cricoid cartilage. They give attachment to the posterior ends of the true vocal cords.

(d) *The Epiglottis:* This is a leaf-shaped plate of yellow elastic fibrocartilage situated in an upright position between the base of the tongue and the upper opening of the larynx. Its lower stalk-like part is attached by a ligament to the thyroid cartilage. The main function of the epiglottis is to prevent food from entering the larynx during swallowing.

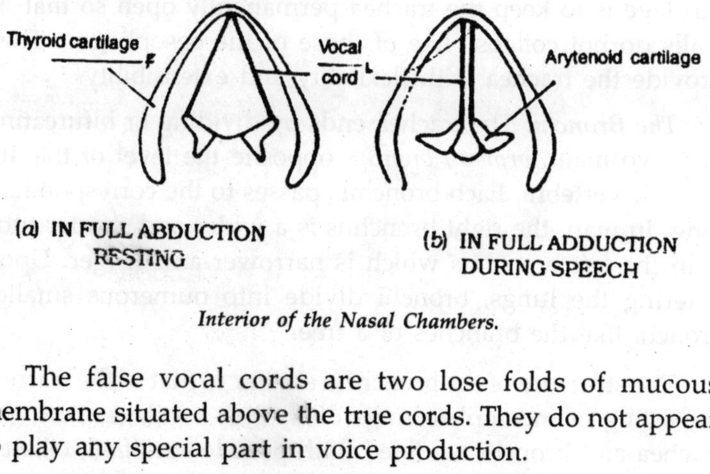

Interior of the Nasal Chambers.

The false vocal cords are two lose folds of mucous membrane situated above the true cords. They do not appear to play any special part in voice production.

Sound has the properties of pitch, loudness and quality.

(i) *The pitch* of the voice depends upon the *length* and *tightness* of the vocal cords. In the male, the cords are longer than in the female (2.3 cm as against 1.7 cm). Thus the male voice has a deeper pitch than the female.

(ii) *The loudness* of the voice depends upon the *force* with which the cords vibrate. The greater the force of expired air, the more vibration of the cords, thus the louder the sound.

The *quality and resonance* of the voice depend upon the shape of the mouth, the position of the tongue, the lips, the factal muscles and on the air sinuses in the bones of the face.

The Trachea: The *trachea* or wind pipe is a continuation of the larynx and extends to approximately are level of the 5th thoracic vertebra where it divides or bifurcates into the left and right primary bronchi. In man the trachea is a relatively thin walled rigid tube, about 40 inches long and 1 inch wide, it lies in front of the oesophagus.

The trachea is composed of from 16 to 20 incomplete, or C shaped rings *of hyaline cartilage,* which are joined together by fibrous and muscular tissue. These rings nearly encircle the trachea, but open posteriorly. The function of the rings of cartilage is to keep the trachea permanently open so that its walls do not collapse like of those of the oesophagus. They provide the trachea with flexibility and extensibility.

The Bronchi: The trachea ends by dividing or bifurcating into two main, *primary bronchi,* opposite the level of the 4th thoracic vertebra. Each bronchus passes to the corresponding lung. In man, the right bronchus is a wider and shorter tube than the left bronchus which is narrower and longer. Upon entering the lungs, bronchi divide into numerous smaller bronchi like the branches of a tree.'

The structure of the bronchi is similar to that of the trachea consisting of incomplete rings of hyaline cartilage. Both the trachea and bronchi are lined with pseudo-stratified, ciliated columnar epithelium containing *goblet cells* which secretes mucous. The mucous helps in holding the dust and bacteria, which are swept towards the pharynx by cilia.

As the bronchi become smaller by subdividing the cartilages become less well defined and more irregular in shape.

The Lungs: The lungs are a pair of conical-shaped organs occupying the greater part of the thoracic cavity. The lungs are separated from each other by *mediastinum* which contains the heart and great vessels, the oesophagus and, in its upper part, the trachea. Each lung is divided by deep fissures into lobes. In rabbit, the left lung is divided into 2 lobes; *left anterior* and *left posterior*, whereas the right lung has 4 lobes; *anterior azygos, right anterior, right posterior* and *posterior azygos*.

In man the left lung has 2 lobes: upper and *lower*. The right lung has 3 lobes: *upper, middle* and *lower*. They extend from the root of the neck above to the diaphragm below. Each lung is described as having a costal and a medial surface, an apex and a base. The *costal surface* is smooth and follows the shape of the chest wall. The *medial surface* of the lung is applied to the mediastinum. Its main feature is the presence of the *hilum* where the primary bronchus and pulmonary artery enter and the pulmonary veins leave the lung.

The *apex* rises into the root of the neck for about one inch above the clavicle. The *base* is concave and is related to the upper surface of the diaphragm.

The Pleura: Each lung is invested by a serous membrane—*the pleura* which consists of two layers: the visceral and parietal.

The Parietal Pleaura: This lines the interior of the chest wall and upper surface of the diaphragm.

The Visceral Pleura: This is firmly attached to the lung itself, completely covering its surfaces, except at the hilum where it is reflected over the structures entering the lung and becomes continuous with the parietal layer.

In normal health, the parietal and the visceral pleura are in *contact* separated only by a thin film of serous fluid. There is a *potential space* between the layers which is termed the *pleural cavity*. The *pleural fluid*, secreted by the epithelial cells of the pleura acts as a lubricant so that the two surfaces can slide smoothly over each other as the lungs expand the contract during the act of breathing.

The Interior of the Lungs: The lungs are composed of the bronchi, bronchioles, alveolar ducts, atria, alveolar sacs and alveoli; of which there are many thousands. The respiratory bronchioles, alveolar ducts, atria, alveolar sacs and alveoli constitute the *lobules* of the lungs. These lobules are separated from one another by the interlobular tissue which is composed of *areolar* and elastic tissue.

Inside the lung, *primary bronchus* divides repeatedly into a number of branches successively of thinner diameter which are called *secondary* and *tertiary bronchi.* These are lined by simple columnar epithelium with abundant cilia and goblet cells. The adventitious layer contains separate cartilaginous plates instead of C-shaped rings. Tertiary bronchi divide into still thinner tubes, the *terminal bronchioles.* In these tubes, an adventitia remains; but cartilage, glands and lymph nodules no longer occur. In the loss of a firm support, the mucosa is thrown into longitudinal folds. Terminal bronchioles are again subdivided in to *respiratory bronchioles.* About two arise from each terminal bronchiole; each subdivides once or twice. A respiratory bronchiole is about 0.5 mm. or less in diameter. In this conduct, *alveoli* appear as little outpouchings that interrupt the main wall at intervals. They project beyond the ordinary wall as still thinner, box-like insets.

Each respiratory branchiole gives rise to several branching *alveolar ducts.* They are relatively long, thin walled fibro-elastic tubes, thickly beset with single alveoli. An alveolar duct terminates into a sort of antechamber, the *atrium.* Varying from 2 to 5 *alveolar sacs* open off from each atrium. Each alveolar sac is composed of many thin-walled, minor compartments, the *alveoli,* all of which open into the main lumen of the sac. An individual alveolus, or air cell, is like a hexagonal box with its top open. In this way, the interior of an alveolar sac presents a honey-combed appearance.

The alveoli have a very thin wall composed of squamous epithelium. It is lined by a layer of mucous and is closely surrounded by a network of capillaries. The blood in capillaries is separated from the air in the alveoli by just two layers of cells, *i. e.*, capillary endothelium and alveolar epithelium. This results in a quick and easy exchange of O_2 and CO_2 between the blood and alveolar air. Each alveolus is thus a miniature lung.

The human lungs have about 750,000,000 alveoli, which have a total surface area of about 100 sq. metres. This comes to about 50 times that of the skin surface.

Mechanical Theory of Respiration

The mechanism of respiration is the process by which the lungs are expanded to take in air, then contracted to expel it. The cycle of respiration consists of 3 phases: (i) *inspiration*, (ii) *expiration* and (iii) *pause.*

This cycle of events occurs about 16 times per minute. The expansion of the chest during inspiration occurs as a result of muscular activity which is partly voluntary and partly involuntary. The main muscles of respiration in normal quite breathing are the *intercostal muscles* and the *diapharagm.*

The Intercostal Muscles: In man there are 11 pairs of intercostal muscles which occupy the spaces between the ribs. They are arranged in 2 layers called: external and internal intercostal muscles.

(1) *The External Intercostal Muscles:* These extend in a *downwards* and *forwards* direction from the lower border of the rib above to the upper border of the rib below. Their action is to *raise the ribs upwards and outwards* thus increasing the size of the thoracic cage *laterally and dorsoventrally.* Because of the increase in the capacity of the thoracic cage, inspiration takes Place.

(2) *The internal Intercostal Muscle:* These extend in a *downwards and backwards* direction from the lower border of the rib above, to the upper border of the rib below, crossing external intercostal muscles at right angles.

These muscles are believed to have an action *antagonistic*

to that of the external intercostal muscles during forced expiration, when they *draw the ribs downwards and inwards.* At the same time, the *abdominal recti muscles* contract forcibly pushing the stomach, liver, etc. upwards against the diaphragm and reducing the volume of the thoracic cavity still further, from that direction.

The intercostal muscles are stimulated to contract by the *intercostal nerves.*

The Diaphragm: The diaphragm is the *main muscle of respiration,* and its action is only partly under control of will. It is a dome-shaped *musculo-fibrous structure* which separates the abdominal and thoraic cavities, thus forming a concave roof for abdominal cavity and a convex floor for the thoracic cavity.

The diapharagm is made up of *radiating muscle fibres* which are inserted at one end into a central *aponeurosis* (tendon) and originate from the circumference of the trunk. When the radiating muscle fibres contract, they pull on the central tendon and *flatten* the dome-shaped structure, thus increasing the *length* of the thoracic cage and, as a result, *inspiration* occurs.

When the muscle of the diaphragm is contracted, it increases the pressure within the abdominal and pelvic cavities, and so assists in *micturition* (the passing of urine) *defaecation* (the passing of faeces) and *parturition* (childbirth).

The diaphragm is supplied by the *phrenic nerves.*

Inspiration: It is described as an *active process* because it is the result of muscle contraction. During *inspiration* (breathing in), the *external intercostal muscles* and the diagphragm contract simultaneously and so the size of the thoracic cage is increased laterally, dorsoventrally and longitudinally at the same time.

As the lungs are held tightly against the thoracic wall. Increase in dimensions of the thoracic cavity results in expansion of the lungs, and pressure within the alveolar and in the air passages is reduced. This results in air being drawn into the lungs in an attempt to equalise the atmospheric and alveolar air pressures.

Expiration: This is a *passive process.* When the external intercostal muscles and diaphragm *relax,* the ribs fall back into place, the diaphragam ascends and again becomes dome-shaped. This is aided by the elastic recoil of lungs an thoraic wall.

With all this, the thoracic cavity decreases in dimensions, permitting the lungs to return to their original size. As they do so, air is forced out of them.

During *vigorous expiration* the antagonistic expiratory muscles—*the internal intercostal muscles,* come into play and decreases the size of the thoracic cavity still further.

After expiration, there is a short *pause.*

Special Respiratory Movements

(1) Cough is a forcible expiration usually proceeded by a prolonged inspiration. The sound of a cough is produced by forcing air though the narrow opening between the vocal cords.

(2) Hiccough is a noisy inspiration caused by muscular spasm of the diaphragm at irregular intervals. The noise is produced by the sudden sucking of air through the vocal cords.

(3) Sighs and yawns are types of prolonged inspiration.

Study of Respiration

Exchange of Gases: Between the process of inspiration and expiration the interchange of respiratory gases occurs. This interchange takes place between the blood in the capillary network which surrounds the alveoli, and the air in the alveoli of the lungs. In external as well as internal respiration, gases always tend to diffuse from a *high partial pressure* to a *lower pressure, i.e..* down the concentration gradient.

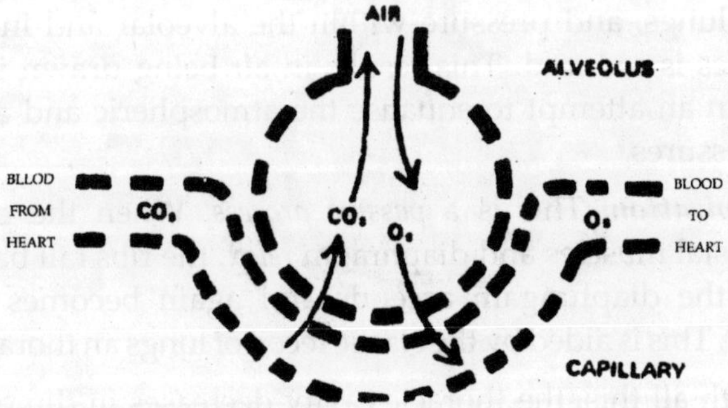

Interchange of gases between air in an alveolus and the blood in a cavillary.

The atmospheric pressure at sea level is 760 millimetres of mercury (mm Hg). This pressure is exerted by the mixture of following gases which make up the air:

Oxygen	20.84%	159.0 mm Hg
Nitrogen	78.62%	597.0
Carbon dioxide	0.04%	0.3
Water vapour	0.50%	3.7
	100%	760.0

Each one of these gases exerts a part of the total pressure depending upon its concentration in the mixture. The proportion of the total pressure provided by each gas is called its *partial pressure.* It is equal to the total pressure of the mixture divided by percentage of that gas in the mixture. For example:
Partial pressure of oxygen = 20.84/100 of 760 = 159.0 mm Hg,
Partial pressure of carbon dioxide = .04/100 of 760 = 0.3 mm Hg

Partial Pressures (mm Hg) of Respiratory Gases

Gas	*Inspired air*	*Alveolar air*	*Venous blood*	*Arterial blood*	*Expired air*
Oxygen	159	104	40	95	120
Carbon dioxide	0.3	40	46	40	27
Nitrogen	597	569	573	573	566

The gas pressure in the blood are usually expressed as gas "tensions", for example, the CO_2 "tension" (PCO_2) is the pressure of the dry gas (mm Hg) with which the dissolved carbonic acid in the blood is in equilibrium; similarly PO_2 (oxygen tension) is the pressure of the dry gas with which the dissolved oxygen in the blood is in equilibrium.

As gases diffuse down the concentration gradient, oxygen diffuses from the alveoli into the blood in the venous capillaries of lungs and carbon dioxide from the blood to the alveoli. This exchange of gases during external respiration is illustrated by the following:

I Oxygen tension in alveolar air =100 mm Hg

Oxygen tension in venous blood = 40 mm Hg

A pressure difference of 60 mm Hg serves to drive oxygen from the alveoli of the lung into the blood.

II CO_2 tension in alveolar air = 40 mm Hg

CO_2 tension in venous blood 46 mm Hg

A relatively small difference of 6 mm Hg is sufficient to drive CO_2 from the blood into the lung. This small difference in pressure is adequate because of the rapidity of the diffusion of CO_2 through the alveolar membrane.

By this mechanism carbon dioxide is excreted from the body and oxygen is absorbed and transported round the body in combination with haemoglobin in the erythrocytes. The blood therefore, now becomes arterial. It has an oxygen tension of about 100 mm Hg and a CO_2 tension of 40 mm Hg.

These gases are dissolved in the blood in simple physical solution. Exchange of gases between the blood and tissue cells during internal respiration takes place again in accordance with the usual physical laws of diffusion. Here, the concentration of O_2 is higher in the blood than in the tissue cells and concentration of CO_2 in greater in the tissue cells than in the blood. Oxygen, therefore, diffuses from the blood into the tissue cells, and CO_2 from tissue cells into blood, at the same time through the interstial (tissue) fluid. Oxygen, thus received by the tissue cells is utilised in oxidative metabolism producing energy and CO_2. This phenomenon maintains oxygen gradient between the blood and tissue cells and the CO_2 gradient between tissue cells and the blood.

The Transport of Oxygen by the Blood: Oxygen is transported in the blood by a respiratory pigment *haemoglobin,* which is a conjugated protein the *chmoprotein.* Haemoglobin consists of a large protein molecule, the *globin* consisting of 4 polypeptide chains (two α and two β chains), to each of which is attached a prosthetic *haem* group.

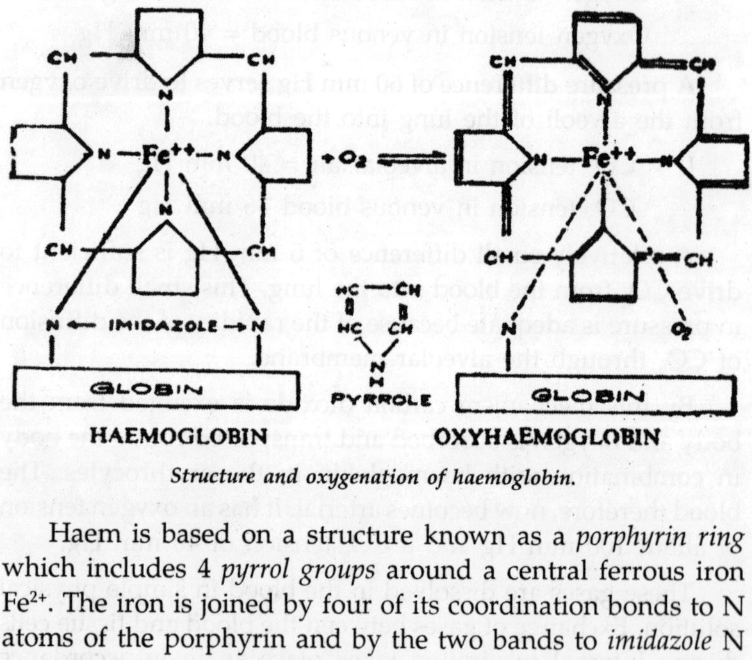

Structure and oxygenation of haemoglobin.

Haem is based on a structure known as a *porphyrin ring* which includes 4 *pyrrol groups* around a central ferrous iron Fe^{2+}. The iron is joined by four of its coordination bonds to N atoms of the porphyrin and by the two bands to *imidazole* N contained in histidine residues within the protein globin (coordination number of iron is 6).

The transport of oxygen by the blood from the lungs to the tissues is due mainly to the ability of haemoglobin to combine *reversibly* with oxygen. When haemoglobin is oxygenated, one of the imidazole N bonds is reversibly displaced by O_2. The reaction is thus a reversible "oxygenation" and not an oxidation. Similarly, reduction is not involved when O_2 is given up. Iron serves as a point of lose physical attachment for O_2 and remains in ferrous (Fe^{2+}) state throughout.

Since one molecule of haemoglobin has 4 haem groups, and each of them is capable of taking up one molecule of O_2; therefore, a mammalian haemoglobin should be depicted as Hb_4O_8.

$$\text{Hb} \quad + 4O_2 \underset{\text{Tissues}}{\overset{\text{Lungs}}{\rightleftharpoons}} \text{-}Hb_4O_8$$

(Deoxygenated or reduced haemoglobin) (Oxyhaemoglobin)

Haemoglobin is *dark red* whereas oxyhaemoglobin is *bright red* in colour.

(i) *Formation of oxyhaemoglobin in Lungs:* This takes place due to three factors:

(a) Normal pH of blood, *i.e.,* 7.4.

(b) Comparatively lower temperature than other parts of the body.

(c) High O_2, low CO_2 concentration (tension).

(ii) *Dissociation of oxyhaemoglobin in tissues:* Three factors account for this process.

(a) Slightly acidic pH due to increase in H^+ concentration: CO_2 produced in the course of metabolism enters the blood where it is hydrated to from carbonic acid, which ionises to form H^+ and HCO_3.

$$CO_2 + H_2O \rightleftharpoons H_2CO_3 \rightleftharpoons H^+ + HCO_3.$$

(b) Rise in temperature.

(c) High CO_2, low O_2 tension.

Oxygen-Haemoglobin Dissociation Curve: The oxygen-haemoglobin dissociation curve shown in fig. demonstrates a progressive increase in the percentage of haemoglobin bound with oxygen as the blood PO_2 increases, which is called the *per cent saturation of the haemoglobin.* As the blood leaving the lungs and entering the systemic arteries usually has a PO_2 of about 95 mm Hg, it can be observed from the dissociation curve that the *usual oxygen saturation of systemic arterial blood is about 97 per cent.* On the other hand, in normal venous blood returning from the peripheral tissues, the PO_2 is about 40 mm Hg, and *the saturation of the haemoglobin is about 75 per cent.* Haemoglobin in the blood automatically delivers O_2 to the tissues at a pressure that is held tightly between about 15 and 45 mm Hg. The curve drawn with CO_2 at a tension of 40 mm Hg is considered as representative of the normal physiologic condition.

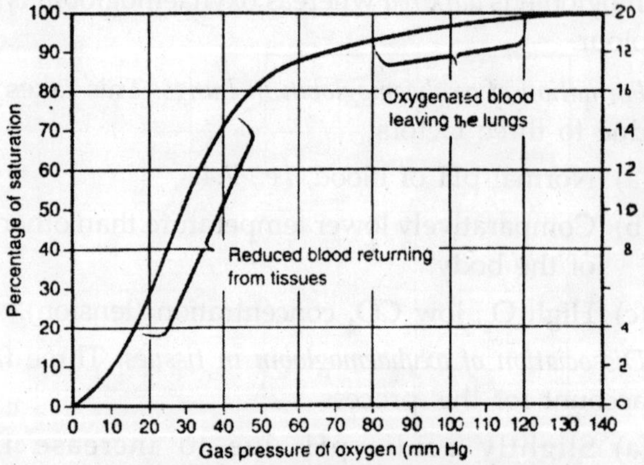

Oxygen-haemoglobin dissociation Curve.

Factors that Shift the Oxygen-Haemoglobin Dissociation Curve: The oxygen-haemoglobin curve of is for normal, average blood. However, a number of factors can displace the dissociation curve in one direction or the other in the manner shown in Fig. This figure shows that when the blood becomes slightly acidic, with the pH decreasing from the normal value of 7.4 to 7.2, the oxygen-haemoglobin dissociation curve shifts, on average about 15 per cent to the right. Conversely, an increase in the pH from the normal 7.4 to 7.6 shifts the curve a similar amount to the left.

In addition to pH changes, several other factors are known to shift the curve. Three of these, all of which shift the curve to the *right, are* (1) increased carbon dioxide concentration (2) increased blood temperature, and (3) increased 2, 3-diphosphoglhycerate (DPG), a metabolically important phosphate compound present in the blood but in different concentration under different metabolic conditions

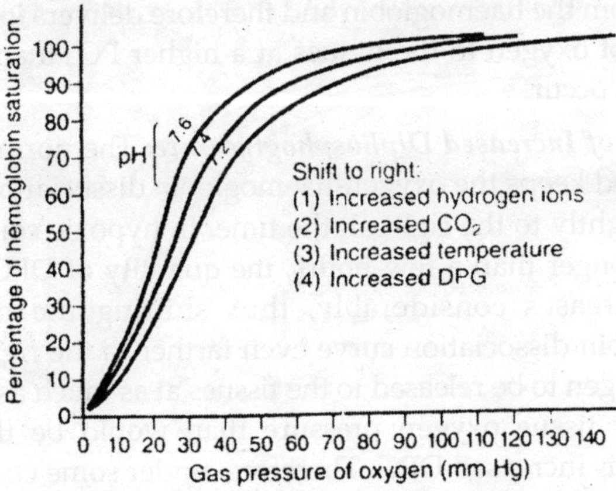

Shift of the oxygen-haemoglobin dissociation curve to the right by increases in (1) hydrogen ions (2) CO_2 (3) temperature, or (4) 2, 3- diphosphoglycerade (DPG).

Increased Delivery of Oxygen to the Tissues When Carbon Dioxide and Hydrogen Ions Shift the Oxygen-Haemoglobin Dissociation Curve—The Bohr Effect. A shift of the oxygen-haemoglobin dissociation curve in response to changes in the blood carbon dioxide and hydrogen ions has a significant effect in enhancing oxygenation of the blood in the lungs and then again in enhancing release of oxygen from the blood in the tissues. This is called the Bohr effect. Which can be explained as follows:

As the blood passes through the lungs, carbondioxide diffuses from the blood into the alveoli. This reduces the blood PCO_2, and decreases the hydrogen ion concentration because of the resulting decrease in blood carbonic acid. Both these effect shift the oxygen-haemoglobin dissociation curve to the left and upward. Therefore, the quantity of oxygen that binds with the haemoglobin at any given alveolar PO_2 now becomes considerably increased, thus allowing greater oxygen transport to the tissues.

Then, when the blood reaches the tissue capillaries, exactly the opposite effects occur. Carbon dioxide entering the blood from the tissues shifts the curve right-ward, which displaces oxygen from the haemoglobin and therefore delivers increased amounts of oxygen to the tissues at a higher PO_2 than would otherwise occur.

Effect of Increased Diphosphoglycerate: The normal DPG in the blood keeps the oxygen-haemoglobin dissociation curve shifted slightly to the right all the time. In hypoxic conditions that last longer than a few hours, the quantity of DPG in the blood increases considerably, thus shifting the oxygen-haemoglobin dissociation curve even farther to the right. This causes oxygen to be released to the tissues at as much as 10 mm Hg higher tissue oxygen pressure than would be the case without this increased DPG. Therefore, under some condition, the DPG mechanism can be important for adaptation to hypoxia, especially to hypoxia caused by poor tissue blood flow.

Respiratory Pigments: In animals, blood usually contains a *respiratory pigment* with a special affinity for oxygen, which increases the oxygen capacity of the blood many times. Four respiratory pigments are found in animals: *haemocyanin, haemerythrin, Chlorocuorin and haemoglobin.* The properties and occurrence of these pigments are summarised in Table.

The Transport of Co_2 in the Blood

CO_2 produced from metabolism in cells diffuses into the blood within the capillaries. It is transported in 2 forms:

(A) The Dissolved State

About 7 per cent of CO_2 combines chemically with water of plasma forming carbonic acid

$$CO_2 + H_2O \rightleftharpoons H_2CO_3$$

Any increase in its concentration causes the dissociation of H_2CO_3 into a hydrogen ion and a bicarbonate ion

$$H_2CO_3 \rightleftharpoons H^+ + HCO_3$$

Transport Mechanism for oxygen within the Blood.

Pigment on oxygenation	*Metal Animal component Group*	*Colour on Occurrence deoxygenation in blood*	*Colour*
Haemocyanin	Copper	Colourless	Blue
Some snails, squids	Plasma		
and Crustacea			
Haemoglobin	Iron	Colourless	Brown
Some polychaeate	Corpuscles		
annelids sipunculids			
and branchiopods			
Chlorocruorin	Iron	Green	
Reddish	Some polychaetes	Plasma	
Some molluscs	Plasma		
Haemoglobin, is a	Iron	Purple-red	
Orange-red	Some annelids	Plasma or	
conjugated protein.	(Ferrous state)		
Corpuscles			
chromo-proteins, which			
Nemertine worms	Corpuscles		
consists of a pigment			
Sea-cucumbes	Corpuscles		
haem (Iron porphyion).			
Fishes	Corpuscles		
attached to the protein,			
Amphibia	Corpuscles		
globin.			
It has the ability to			
Reptiles	Corpuscles		
combine reversibly			
Birds	Corpuscles		
with oxygen			
Mol. Wl. 6. 450.			
Mammals	Corpuscles		

As Chemical Compounds

(1) *Carbamino Compounds:* About 23 per cent of the total blood CO_2 is transported in combination with haemoglobin.

$$CO_2 + Hgb.\,NH_2 \longrightarrow Hgb.\,CO_2$$

(Free amino-group of haemoglobin) (Carbamino haemoglobin)

(2) *Bicarbonates:* About 70 per cent of CO_2 produced as a result of oxidative metabolism in the body is carried as bicarbonates in the blood.

(i) *In the Erythrocytes:* CO_2 from the plasma enters in red blood cells and combines with water within the cells. The reaction is catalysed by a zinc containing enzyme *carbonic anhydrase,* which is associated with the haemoglobin in red-cells (never in the plasma). *Carbonic anhydrase* catalyses the reaction between CO_2 and water, accelerating its reaction rate about 5000 fold. Instead of requiring many seconds or minutes to occur, as in the case of plasma, the reaction occurs so rapidly in the red blood cells that it reaches almost complete equilibrium within a very small fraction of a second. This results in the formation of carbonic acid which soon dissociates:

$$CO_2 + H_2O \xrightarrow{\text{Carbonic anhydrase}} H_2CO_3 \rightleftharpoons H^+ + HCO_3^-$$

A small amount of HCO_3^- is transported in the erythrocytes.

In the red-cells, haemoglobin remains combined with potassium and forms bicarbonate in the following way:

$$H^+ + HCO^-_3 + K.Hb \longrightarrow KHCO_3^- + H.Hb$$

(Potassium Haemoglobinate) (Haemoglobin acid)

The hydrogen ions mostly combine with Hb for keeping the pH of blood (7.4) in steady state, because Hb is a powerful acid-base buffer.

(ii) *In the Plasma:* Most of the HCO_3^- diffuse into the plasma. Plasma transports CO_2 by three different mechanisms.

(a) *By phosphate Buffers:* Alkaline phosphates

combine with carbonic acid in the plasma and form sodium bicarbonates.

$$Na_2HPO_4 + H_2CO_3 \longrightarrow NaHCO_3 + NaH_2PO_4$$

(Di-sodium hydrogen phosphate) (Sodium dihydrogen phosphate)

(b) By *plasma Proteins:* Proteins of all plasma mostly remain combined with Na and form bicarbonates in the following manner:

$$NaPr + H_2CO_3 \longrightarrow H.Pr + NaHCO_3$$

(Sodium proteinate) (Proteinic acid)

(c) *By Chloride shift or Hamburger's Phenomenon:* CO_2 released from tissue metabolism enters the red cell (erythrocyte).

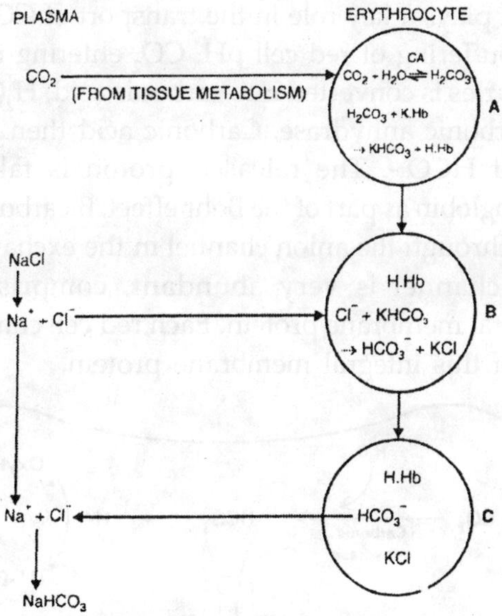

Under normal conditions the walls of the red cells are permeable to anions (Cl, HCO_3) but virtually impermeable to cations (Na^+, K^+) and to the large haemoglobin molecule. Under these circumstances chloride ions diffuse into the erythrocyte from the plasma and react with intracellular potassium bicarbonate. The bicarbonate ions which are thus released, now diffuse out of the erythrocyte into the plasma, whereas chloride ions are neutralised by the potassium in the erythrocyte. In the plasma, bicarbonate ions combine with sodium ions forming sodium bicarbonate ($NaHCO_3$). This phenomenon is known as chloride shift which permits the carriage of additional CO_2 as sodium bicarbonate by plasma.

All the reaction of chloride shift are reversible. At the lung, when the blood becomes arterial, chloride shifts back into the Plasma, thus, liberating intracellular potassium to buffer the newly formed oxyhaemoglobin and, in the plasma, neutralising the sodium liberated by the removal of CO_2 during respiration.

Anion Channel: The red-cell membrane contains an anion channel that plays a key role in the transport of CO_2 by blood and in the buffering of red cell pH. CO_2 entering red cells in tissue capillaries is converted into carbonic acid, H_2CO_3, by the action of carbonic anhydrase. Carbonic acid then dissociates into H^+ and HCO_3-. The released proton is taken up by deoxyhaemoglobin as part of the Bohr effect. Bicarbonate leaves the red cell through the anion channel in the exchange for Cl^-. The anion channel is very abundant, comprising about 1/3 of the total membrane protein. Each red cell contains some 10^6 copies of this integral membrane protein.

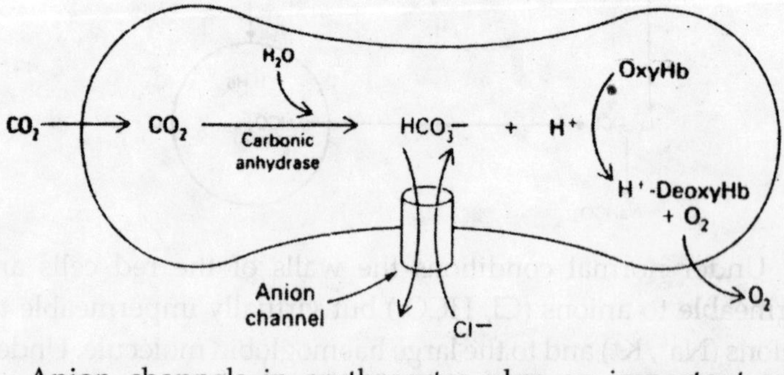

Anion channels in erythrocytes play an important role in the transport of CO^2 by enabling HCO^{3-} to be exchanged for Cl^-.

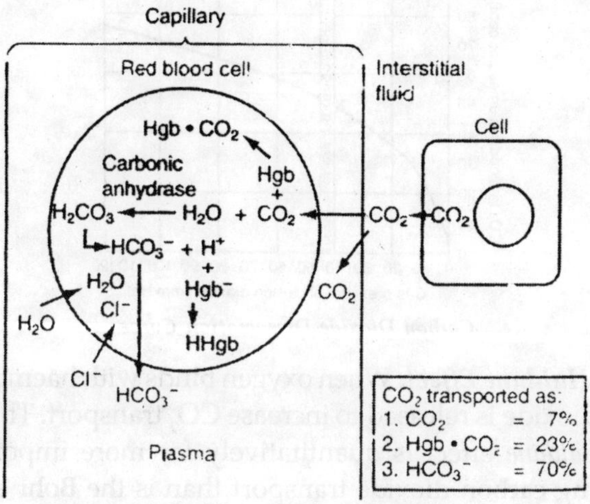

Release of Co_2 at the Respiratory Surface: Carbonic acid, bicarbonates of sodium and potassium and car-bamino-compounds are carried to the lungs where they break down under the influence of various factors and liberate free CO_2.

(i) $H_2CO_3 \xrightarrow[\text{anhydrase}]{\text{carbonic}} H_2O + CO_2\uparrow$

(ii) $2KHCO_3 \xrightarrow{\text{Acidity of Oxyhaemoglobin}} K_2CO_3 + H_2O + CO_2\uparrow$

(iii) $2NaHCO_3 \xrightarrow{\text{Acidity of Oxyhaemoglobin}} Na_2CO_3 + H_2O + CO_2\uparrow$

(iv) $Hb.NH.\ COOH \xrightarrow[\text{Concentration of } O_2]{\text{Reduction due to higher}} Hb.\ NH_2 + CO_2\uparrow$

Carbon Dioxide Dissociation Curve: The normal blood PCO_2 ranges between the limits of 40 mm Hg in arterial blood and 45 mm Hg in venous blood, which is a narrow range. The normal concentration of carbon dioxide in the blood in all its different forms is about 50 volumes per cent but only 4 volumes per cent of this is exchanged during normal transport of carbon dioxide from the tissue; to the lungs. That is, the concentration rises to about 52 volumes per cent as the blood passes through the tissues and falls to about 48 volumes per cent as it passes through the lungs.

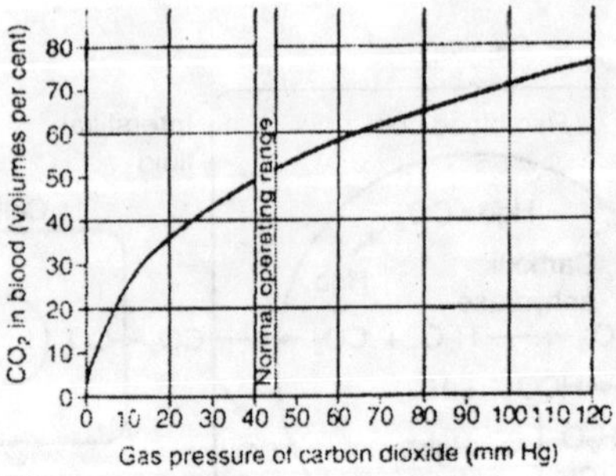

Carbon Dioxide Dissociation Curve

The Haldane Effect: When oxygen binds with haemoglobin, carbon dioxide is released to increase CO_2 transport. This effect called *Haldane* effect is quantitatively far more important in promoting carbon dioxide transport than is the Bohr effect in promoting oxygen transport.

The Haldane effect results from the simple fact that combination of oxygen with haemoglobin in the lungs causes the haemoglobin to become a stronger acid. This in turn displaces carbon dioxide from the blood and into the alveoli.

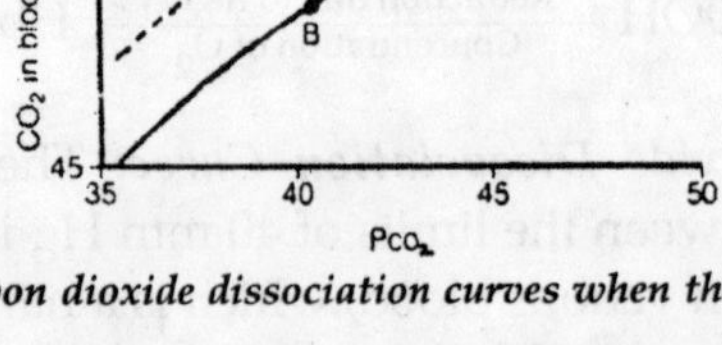

Portions of the carbon dioxide dissociation curves when the PO_2 is 100 mm Hg.

In the tissue capillaries, the Haldane effect causes increased pickup of carbon dioxide because of oxygen removal from the haemoglobin and in the lungs, it causes increased release of carbon dioxide because of oxygen pickup by the haemoglobin.

Fig. demonstrates quantitatively the significance of the Haldane effect on the transport of carbon dioxide from the tissues to the lungs. This figure shows small portions of two carbon dioxide dissociation curves, the solid curve when the PO_2 is 100 mm Hg, which is the case in the blood capillaries of the lungs, and the dashed curve when the PO_2 is 40 mm Hg, which is the case in the tissue capillaries. Point A on the dashed; curve shows that the normal PCO_2 of 45 mm Hg in the tissues causes 52 volumes per cent of carbon dioxide to combine with the blood. On entering the lungs, the PCO_2 falls to 40 mm Hg while the PO_2 rises to 100 mm Hg.

If the carbon dioxide dissociation curve did not shift because of the Haldane effect, the carbon dioxide content of the blood would fall only to 50 volumes per cent, which would be a loss of only 2 volumes per cent of carbon dioxide. However, the increase in PO_2 in the lungs decreases the carbon dioxide dissociation curve from the dashed curve to the solid curve of the Fig. so that the carbon dioxide content falls to 48 volumes per cent (point B).

This represents an additional 2 volumes per cent loss of carbon dioxide. Thus, the Haldane effect approximately doubles the amount of carbon dioxide released from the blood in the lungs and approximately doubles the pickup of carbon dioxide in the tissues.

Carbon Monoxide (Co) Poisoning: Carbon monoxide poisoning may occur under any circumstances where carbon-containing compounds [*e.g.*, coal, gas, oil, etc.) are burned with deficient oxygen supply.

In the presence of sufficient oxygen,

$$C + O_2 \longrightarrow CO_2$$

But if oxygen is deficient,

$$2C + O_2 \longrightarrow 2CO_2$$

The carbon monoxide poisoning cases occur when a person sleeps in a closed room with a burning coal-stove on a winter night, or when a person remains inside a closed motor garage for a long time with a running automobile engine. As there is insufficient oxygen in such a closed environment, CO is produced from the burning stove or from the running motor engine (exhaust gas) and proves fatal to the inhabitant.

In CO poisoning, haemoglobin combines with carbon monoxide even more readily than with oxygen (230 times as fast), to form a cherry-red compound *carboxyhaemoglobin.* This reduces the amount of haemoglobin available to carry oxygen. Moreover, CO is not easily released by the haemoglobin. When the CO in the inspired air is as low as 0.02 per cent, headache and nausea will occur. If the CO concentration is only 1/230 that of oxygen in the air (approximately 0.1 per cent CO), unconsciousness will occur in 1 hour and death in 4 hours. The skin of the victim of CO poisoning develops a cherry-red colour because this is the colour which haemoglobin attains when it combines with CO.

The treatment of CO poisoning is to bring about the break-down of HCO, replacing CO with oxygen. The application of O_2 and prompt removal of the victim from the fouled atmospherically save his life. To stimulate the respiratory centre of the victim to greater effort, 6 per cent, CO_2 is added to the O_2.

Subdivisions of Air in the Lungs

Pulmonary Volumes/Aerodynamics: It is customary to divide the volume of air entering the lungs into various sub-division. These are as follows:

(i) *Tidal Volume:* The air normally inspired and expired at one breath is called the tidal volume. It amounts to about 500 ml.

(ii) *Complemental Air or Inspiratory Reserve Volume (IRV):* After a normal inspiration, it is still possible to send more air into the lungs by forcible inhalation. The air inhaled in this manner is called the complemental air (3000 ml).

(iii) *Supplemental Air or Expiratory Reserve Volume (ERV):* After a normal expiration, it is possible to expel more air by breathing out forcibly. The additional air pushed out like this is called the supplemental air (1100 ml).

(iv) *Vital Capacity (VC):* The amount of air breathed in and out with the greatest possible effort is said to be the vital capacity of the lungs. It is equal to the sum of the tidal complemental and supplemental air (0.5 + 3.0 + 1.1 = 4.6 L).

(v) *Residual Air or Residual Volume (RV):* Some air is left in the lungs after the most forceful expiration. This air is known as the residual air (1200 ml) and can be released only by opening the thoracic cavity.

(vi) *Dead Air Space:* The only place in the lungs where gaseous exchange takes place is in the alveoli. A part of the inspired air is left in the trachea and bronchial tree, the so-called dead air space, where no gaseous exchange occurs. This dead space air (about 150 c.c) is expelled at the next expiration. The expired air, therefore, contains fresh air from the dead space and foul air from the lungs.

	Volume (L)	
Tidal volume	0.5	quite breathing
Inspiratory reserve volume	3.0	forced inspiration
Expiratory reserve volume	1.1	forced expiration
Vital capacity	4.6	
Residual volume	1.2	always left in lung
Total lung capacity (TLC)	5.8	

(vii) *Total Lung Capacity (TLC):* It can be seen from the above table that the total lung capacity (TLC) is the sum of the vital capacity (VC) and the residual volume (RV), *i.e.*,

$$TLC = VC + RV$$

(viii) *Pulmonary Ventillation:* The amount of air inspired per minute is called pulmonary ventillation. It is normally about 6L (500 mL/breath × 12 breaths/min).

(ix) *Maximal Voluntary Ventilation (MW):* It is the *maximal breathing capacity* or the largest volume of gas that can be moved into and out of the lungs in 1 minute by voluntary effort. The normal MW is 125—170 L/min.

Respiratory Quotient: The amount of oxygen consumed and the amount of carbon dioxide given off during respiration, are often used as a measure of the extent of oxidation of fat, carbohydrate and protein by a particular animal. For example, in the oxidation of a hexose sugar the quantity of carbon dioxide formed is equal to the quantity of oxygen utilised:

$$C_6H_{12}O_6 + 6CO_2 \rightleftharpoons 6O_2 + 6H_2O$$

Here the ratio is $\dfrac{\text{volume of } CO_2 \text{ formed}}{\text{volume of } CO_2 \text{ utilized}}$ equal to unity.

This ratio, which is known as the *respiratory quotient* (RQ) is different for each of the three major food stuffs. In the case of a fat such as triolein, the equation for its oxidation is as follows:

$$C_{57}H_{104}O_6 + 80O_2 \longrightarrow 57CO_2 + 52H_2O$$

In this case, the RQ is therefore equal to 57/80, or 0.71. Protein is not completely oxidised in the tissues, but its RQ can be determined by indirect means, and shown to be between 0.8 and 0.82.

Reasonable Respiration

The control of respiration is partly *nervous* and partly *chemical.* The two modes are closely linked to each other

Nervous Control: Located bilaterally in the medulla oblongata and pons varolii of the brain is a collection of highly specialised and widely dispersed groups of nerve cells collectively called the respiratory centre. From this centre, nerve impulses pass out in the *phernic nerves* to the *diaphragm* and *interocostal nerves* to the *intercostal muscles.* These impulses stimulate the muscles concerned to contract and increase the capacity of the throacic cavity with the result that inspiration occurs.

Chemical Control: The main chemical factors which influence the respiratory centre are *the amount of carbon dioxide (carbonic acid) and oxygen in solution in the blood.* In the carotid body, where the common carotid artery divides, and in the wall of the arch of the aorta there are cells which are sensitive to carbon dioxide excess and *oxygen lack. When these cells are stimulated, impulses pass in branches of the vagus and glossopharyngeal nerves to the respiratory centre* in the *medulla oblongata.*

The accumulation of carbon dioxide in solution in the blood is the most important stimulus to respiration. If the amount of carbon dioxide rises in the blood. For example during muscular exercise, the respiratory centre is stimulated. It therefore sends out impulses to the respiratory muscles to produce deeper and quicker breathing so that the carbon dioxide can be given out more rapidly by the lungs. In this way, the respiratory centre controls the rate and depth of breathing.

Physiology of Respiration: All living animals require some energy to perform their activities. The most common source of energy is oxidation of food by the available oxygen that is respired by the animal. *Thus, respiration can be defined as a catabolic process in which the respired oxygen is used in the oxidation of food resulting in the release of energy.* Respiration, therefore, includes both breathing and oxidation. However, the process of taking in oxygen and giving off carbon dioxide is also, sometimes, called respiration. But this is only the breathing process and does not take into consideration the oxidation process, without which the main purpose of respiration cannot be achieved.

In respiration, the animals make use either of the oxygen of the atmospheric air (wherein it is about 21 per cent of the air), or of the oxygen dissolved in the water.

There are two kinds of respiration.

Direct Respiration: Direct respiration is the exchange of oxygen and carbon dioxide between the cells of the organism and the surrounding medium. For example, small aquatic animals take up oxygen by diffusion from the small amount of this gas dissolved in water.

Indirect Respiration: Indirect respiration means the exchange of gases through some specialised structure (often called the respiratory organ) and involves two phases, external and internal respiration.

(a) *External Respiration:* External respiration includes the exchange of gases between the external environment and the blood of the respiratory organ.

(b) *Internal Respiration:* Internal respiration, or cellular respirations involves the exchange of gases between the bloodstream and the body cells. It is also known as *tissue respiration* and the oxygen consumption (microlitres O_2) per milligram of tissue per hour is denoted as Qo_2. Since the respiratory rate of each tissue depends on its demand for ATP, there are considerable variations in Qo_2 of various tissues of the body.

Internal respiration fulfils the ultimate object of the respiratory system in that it provides oxygen to cells and tissues of the body and rids them of carbon dioxide.

Significance of Respiration in Invertebrates

Invertebrates depend upon the following modes of respiration:

(a) *Direct Diffusion:* In protozoa, porifera, coelenterata and many worms oxygen is obtained by diffusion directly from the surrounding medium (water).

(b) *Integument:* In most annelids, gaseous exchange is made through a moist skin provided with blood vessels or other derivatives of the integument like parapodia which are highly vascular. A few crustaceans and molluscs also use their skin for respiration. In some molluscs *(e.g., Pila)* mantle is used and in *Palaemon,* there are three pairs of epipodites which function as respiratory organs. Some echinoderms make use of the finger-like dermal branchiae (papulae), as well as the tube feet.

(c) *Gills*: Gills are common in aquatic invertebrates. Some are simple in structure (as in marine annelids) and others are complicated (as in molluscs and arthropods). Numerous aquatic insects have gills through which oxygen passes directly to the tracheae.

(d) *Tracheae:* These are found in onychophora, insecta, myriapoda, and in some spiders and isopods. Tracheae usually open by spiracles that can be closed or opened according to the need.

(e) *Lungs:* The so-called lungs of invertebrates are a variety of structures. In land snails, mantle is modified as lung. Scorpions and spiders have booklungs opening on the ventral surface through small slit-like apertures called the stigmata.

(f) *Respiratory Trees:* Respiratory trees of holothurians are filled with sea water by cloacal pumping and contractions of the trees themselves, and gas exchanges with the coelomic fluid.

Significance of Respiration in Vertebrates

In vertebrates, the respiration is carried out by any of the following three devices.

(a) *Direct Surfaces such as the Skin and Pharyngeal Regions*: Many amphibians make use of their skin in respiration. Some salamanders depend on the skin and highly vascular pharyngeal region. Some fishes and eels also absorb oxygen through the skin.

(b) *Gills:* Many aquatic vertebrates have gills, either external (protochordates and some larval amphibians) or internal (cyclostomes and fishes).

(c) *Lungs:* Land vertebrates have the lungs for respiration. In amphibians, the lungs are simple ; but in higher vertebrates there is a gradual increase in the internal respiratory surface of the lungs, thus making them more efficient organs of respiration.

Respiration in Humans and other Mammals

Respiratory System: It consists of certain air channels and the lungs. The respiratory passage consists of nostrils (external nares), nasal chamber, posterior nares, pharynx, epiglottis, glottis, larynx or voice box, trachea or windpipe and two bronchi. Within the lungs each bronchus divides and subdivides into smaller tubes (bronchioles) which lead to the *air sacs* and alveoli.

The human lungs are made up of some 7.5×10^8 alveoli, the total surface of which is about 100 square metres, or about fifty times that of the skin surface. Thus, each lung consists of its bronchial tree with its many air sacs and alveoli units together with other asscciated structures such as the blood vessels, nerves and pleura. All are supported and attached to one another by the connective tissue. Externally they are covered by the *visceral pleura* (a thin layer of smooth epithelium) which is also surrounded by another similar layer, the *parietal pleura* forming the inner lining of the wall of the chest.

A small amount of pleural fluid for lubrication is found in the *pleural cavity.* The normal pressure of the fluid in the intrapleural space is -10 to -15 mm Hg. This unusually negative pressure in the intrathoracic space is caused by the continuous reabsorption of fluid into the pleural capillaries by a special lymph-pump system. The *diaphragm,* a dome-shaped partition forms the floor of the chest cavity and separates it from the abdomen.

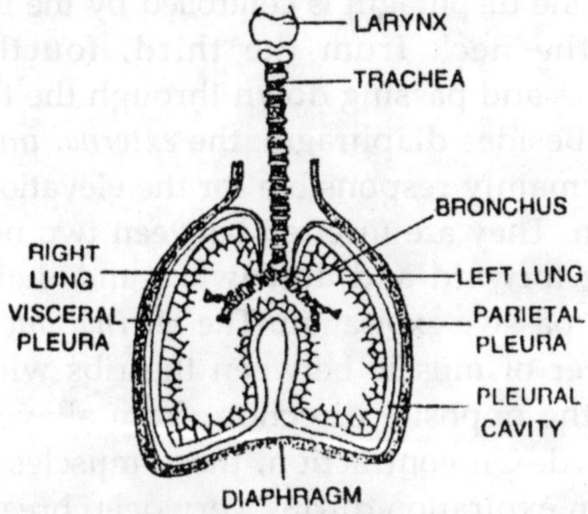

Lungs and diaphragm in man.

Breathing Mechanism

Breathing is a mechanical process and is completed in two phases, inspiration and expiration. In *inspiration,* the ribs are elevated and the diaphragm contracted and flattened, the chest cavity is enlarged. This increase in the volume of the chest cavity and lungs causes the air pressure in the lungs to fall below the atmospheric pressure, and air passes through the air passageways to the lungs to equalise the pressure. In *expiration,* the ribs and diaphragm return to their original position, the chest cavity volume is decreased. The distended elastic lungs then contract and the air is forced out. Changes in the intrapleural pressure are responsible for air entering and leaving the lungs. In inspiration, expansion of the thorax, aided by descent of the diaphragm, decreases intrathoracic pressure from -4 to -10 mm Hg, and air pushes into the lungs. Thus, in inspiration the lungs are extending passively in response to the various mechanisms that result in an increase in thoracic volume. In expiration, the size of the thorax is decreased, the intrathoracic pressure is raised to-2 mm Hg, and air is forced out of the lungs.

It should be noted that the diaphragm is the main muscle of inspiration. If the diaphragm descends 10 mm, it will increase the thoracic cavity volume by 250 ml. When it relaxes, passive expiration results. The contraction and relaxation of the diaphragm is controlled by the *phrenic nerves* arising in the neck from the third, fourth and fifth cervical nerves and passing down through the thorax to the diaphragm. Besides diaphragm, the *external intercostals* are the muscles mainly responsible for the elevation of the ribs in inspiration. They are inserted between two neighbouring ribs, sloping forward and downward and their relaxation brings about passive expiration. The *internal intercostals* form a deeper layer of muscle between the ribs with the fibres running in the opposite direction, from above downward and backward. On contraction, these muscles depress the ribs aiding in expiration during very deep breathing (active expiration).

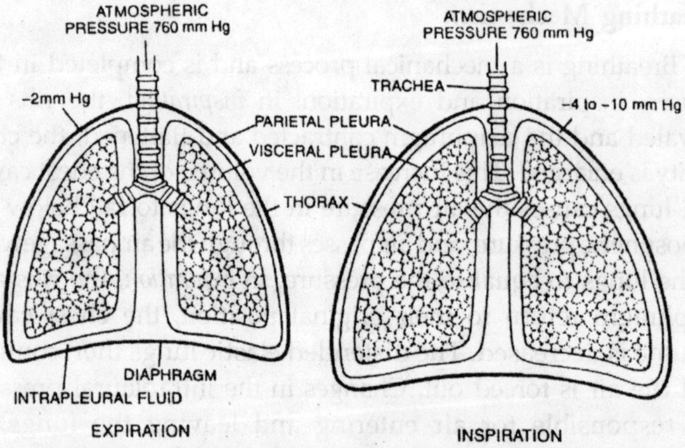

Showing changes in intrapleural pressure responsible for inspiration and expiration.

Generally, the following terms are used to describe respiration.

(a) *Eupnea* : normal respiration

(b) *Hypernea* : increase in respiratory rate and depth

(c) *Dyspnea* : irregularities of respiration

(d) *Apnea* : cessation of respiration.

The normal rate of respiration in the adult is 14 breaths/minute, but in children it may be up to 30/minute. In exercise it is further increased. Each inspiration admits about 350 ml of new air to mix with the 2500 ml of old air present in the lungs. The quantity of new air that enters the lungs per minute is known as the *minute volume,* which in the average adult is about 4900 ml (350 × 14).

Physiology of Respiration: It may be discussed under the following heads:

(a) Gaseous exchange in the lungs; (b) Transport of gases by the blood; and (c) Cellular (internal) respiration and oxidation.

Gaseous Exchange in the Lungs (external respiration): The respiratory membrane through which gaseous exchange takes place, consists of the thin lining of the alveoli, the endothelium of the capillaries, and the delicate interstitial connective tissue layer. The uptake of oxygen and the release of carbon dioxide by the blood of the alveolar capillaries can be explained by *diffusion, i.e.,* the gases pass from the regions of high pressure to those of low pressure.

The pressure of the gas refers to the *partial pressure* that the gas exerts in a mixture of gases. If the atmospheric pressure at the sea level is 760 mm of mercury, the partial pressure of oxygen will be 21 per cent (percentage of O_2 in air) of 760 mm or 159 mm. The partial pressure of O_2 (PO_2) in the alveolar air is about 100 mm of mercury (Since O_2 constitutes only 13 per cent of alveolar air) compared with that of 159 mm of mercury for the O_2 in the atmosphere. The partial pressure of O_2 in the blood of the alveolar capillaries is considerably lower, being only about 40 mm. Accordingly, the oxygen diffuses from the alveolar air into the blood. By the time the blood leaves the capillaries, its O_2 pressure has been raised to about 100 mm of Hg, the same as alveolar O_2.

In a similar manner, the CO_2 in the blood of the lung capillaries has a higher concentration (46 mm) than it has in the lung alveoli (40 mm). This accounts for the diffusion of CO_2 out of the blood into the alveoli. By the time the blood leaves the lungs, its CO_2 pressure has been lowered to approximately

40 mm. Thus, in the diffusion of both these gases there is a marked *diffusion gradient* that determines the direction of their flow.

The exchange of gases in the lung according to their partial pressures (concentration gradient) may be shown as follows:

Lung alveoli	100 mm Hg	40 mm Hg
	↓ O_2	↑ CO_2
Lung capillaries	40	46

The large volume of *inert gases* (mainly nitrogen), that is inspired normally passes into the blood with oxygen and carbon dioxide. As these gases are *chemically inactive,* they do not participate in the metabolic reactions of the body.

Blood as a Gas Carrier

Transport of Oxygen: The solubility of O_2 in water is rather low, but this shortcoming is overcome by the fact that the O_2 is bound to carrier substances in the blood. In mammalian blood, the O_2 carrier respiratory pigment is *haemoglobin.* The maximum amount of O_2 which the normal human blood can absorb is 20 ml per 100 ml of blood. When O_2 passes from the lung alveoli into the lung capillaries, it diffuses into the blood and unites with haemoglobin to form *oxyhaemoglobin.*

$$Hb_4 + 4O_2 \rightleftharpoons Hb_4O_8 \text{ or } Hb_4\,(O_2)_4 \text{ (Oxyhaemoglobin)}$$

As each haemoglobin molecule possesses 4 atoms of iron, it can transport 4 molecules of oxygen.

Under the normal conditions the arterial blood which has been exposed to the alveoli of the lungs is not quite completely oxygenated. With an O_2 tension of 100 mm of Hg, it is usually 98 per cent saturated and therefore, contains 19.6 ml of O_2 (combined to haemoglobin) per 100 ml of blood. In addition to this there is about 0.2 to 0.3 ml of O_2 which is dissolved in the plasma.

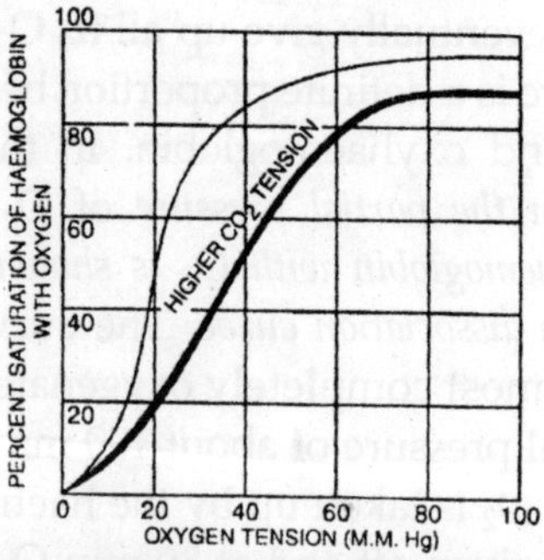

The oxygen-haemoglobin dissociation curve. Increased CO_2 concentration shifts the curve to the right.

The arterial blood and the alveoli have the same O_2 pressure (100 mm of Hg). But in the cells and tissues of the body the O_2 tension is considerably low (1 to 40 mm of Hg). The O_2 is accordingly liberated from the oxyhaemoglobin and diffuses out from the blood through the thin capillary walls into the cells. This is made possible by the important fact that the combination between O_2 and haemoglobin in the red blood cells to form oxyhaemoglobin is a *reversible* one. The liberation of O_2 from the blood to the tissues is just as important as its rapid absorption by the blood during its passage through the lungs.

$$\underset{\text{Oxyhaemoglobin}}{Hb_4O_8} \longrightarrow \underset{\text{reduced haemoglobin}}{4Hb + 4O_2}$$

The reduced haemoglobin is further transported via blood to the lungs and the cycle is repeated, while the O_2 that has diffused into the cells is utilised in the oxidation of carbohydrates, resulting in the release of CO_2 and energy.

Graph Representing Dissociation of Oxygen and Haemoglobin

At high O_2 pressure, the haemoglobin combines with O_2 to form oxyhaemoglobin. Each iron atom can bind one O_2 molecule, and when all sites are occupied, the haemoglobin cannot take on any more, since it is fully loaded or *saturated.* At low O_2 pressure, O_2 dissociates from its binding, and the haemoglobin will eventually give up all its O_2. At any given O_2 concentration, there is a definite proportion between the amount of haemoglobin and oxyhaemoglobin. In this way the *actual relationship between the partial pressure* of O_2 *and the degree of saturation of the haemoglobin with* O_2 *is shown by the remarkable oxygen haemoglobin dissociation curve.* The curve shows that the haemoglobin is almost completely oxygenated (saturated) with O_2 at the O_2 partial pressure of about 100 mm Hg. At higher O_2 pressure, no more O_2 is taken up by the haemoglobin. At lower O_2 pressure, O_2 is given off and at 30 mm O_2 pressure, half the haemoglobin is present as oxyhaemoglobin. As the O_2 pressure decreases further, more O_2 is given off, and all is given up when the O_2 pressure reaches zero. Thus, the degree of haemoglobin saturation is lowered with the fall in the partial pressure of O_2. In the passage of blood through the tissues where the O_2 tension is low, rapid dissociation of oxyhaemoglobin occurs, yielding a comparatively large quantity of O_2 to the surrounding tissues and cells where it is most needed.

Lungs (P_{o2} = 100 mm Hg)

$$Hb + O_2 \longrightarrow HbO_2 \text{ (oxyhaemoglobin)}$$

Tissues (P_{o2} = 30 to 40 mm Hg)

$$HbO_2 \longrightarrow Hb + O_2$$

During exercise, there is a fall in tissue PO_2, an increase in Pco_2, and an increase in pH, local temperature and 2, 3-diphosphoglycerate concentration. All these factors promote the release of oxygen from oxyhaemoglobin (shifting the oxygen-haemoglobin dissociation curve to the right) and thus increasing the efficiency of oxygen delivery to the active tissues.

Factors Affecting the Oxygen-Haemoglobin Dissociation Curve

(i) *Effect of Partial Pressure of* O_2 : The degree of saturation of haemoglobin completely depends on the partial pressure of O_2. An examination of the curve tells us that haemoglobin is almost completely oxygenated by an O_2 tension of 100 mm Hg, which is the same as that in the lungs. Therefore, an increase in the partial pressure of O_2 beyond 100 mm Hg would have nothing to do with the saturation of haemoglobin with O_2. However, the degree of haemoglobin saturation is lowered with the fall in the partial pressure of O_2. This explains the dissociation of oxyhaemoglobin at low partial pressure of O_2.

(ii) *Effect of Partial Pressure of CO_2:* The oxygen dissociation curve is also influenced by CO_2. An increase in the CO_2 shifts the curve to the right, a displacement that is called the *Bohr effect* after the name of its discoverer. A higher CO_2 concentration, therefore, causes more O_2 to be given up at any given O_2 pressure, a result that is of significance in the tissues, for as the blood in the capillary gives off oxygen, CO_2 enters from the tissues and increases the amount of O_2 that is given off from the haemoglobin. Thus, *Bohr effect* facilitates the delivery of additional O_2 from the blood to the tissues. Conversely, the uptake of O_2 and the formation of oxyhaemoglobin in the blood coursing through the lungs greatly reduces the carbon dioxide-carrying capacity of the blood, thus facilitating the release of CO_2 to the alveoli.

(iii) *Effect of Temperature:* A rise in temperature also causes a decrease in the oxygen-carrying capacity of the blood. For example, during increased muscle activity, the rise in acidity due to CO_2 and lactic acid production and the localised increase in temperature results in a greater liberation of O_2 to the tissues. Thus, temperature affects the dissociation curve of O_2 by affecting the partial pressure of O_2.

(iv) *Effect of Haemoglobin:* It is apparent that the quantity of O_2 transported in the blood will depend on the haemoglobin content of the blood. Blood containing less haemoglobin (as in anaemia) carries less O_2 to the cells and tissues.

(v) *Effect of pH:* The acidity of the blood (pH) also affects the degree of haemoglobin saturation with oxygen. An increase in the metabolic rate in the tissues increases the CO_2 and acid metabolites. The O_2 partial pressure is thus lowered, thereby more O_2 is liberated and made available to the tissues.

(vi) *Effect of 2, 3- diphosphoglycerate (DPG):* A major fraction of the phosphate in erythrocytes is present as DPG which plays an important role in the release of oxygen to the tissues. An increase in DPG concentration is associated with decreased oxygen levels such as are found in anaemia or cardiac inefficiency. DPG reacts with haemoglobin to reduce the affinity of haemoglobin for oxygen and so makes more oxygen available to the anoxic tissues. In some genetic abnormalities, DPG is reduced and the red blood cells have a short survival time.

$$HbO_2 + DPG \rightleftharpoons Hb.\ DPG + O_2$$

Transport of Carbon Dioxide: Carbon dioxide is evolved in the body as a result of various metabolic activities of the cells and diffuses into blood. The total amount of CO_2 in the venous blood is about 60 ml per 100 ml blood, and the arterial blood contains about 50 ml total CO_2 per 100 ml. Thus, a relatively small amount of CO_2 is given off in the lungs. Carbon dioxide, that diffuses into the blood, is transported in the following three ways:

Transport of CO_2 in Physical Solution: As CO_2 enters the blood from the tissues, it combines with water of the plasma to form carbonic acid (H_2CO_3). Thus about 5 per cent of CO_2 is carried in solution in the plasma as carbonic acid.

$$CO_2 + H_2O \rightleftharpoons H_2\ CO_3 \rightleftharpoons HCO_3^- + H^+$$

These ions then combine with the buffers of the blood.

Transport of CO_2 as Carbamino Compounds: About 10 per cent of CO_2 is transported as carbamino compounds. In the red blood cells it combines directly with the amino groups (—NH_2) of the haemoglobin to form the so-called carbohaemoglobin.

$$CO_2 + Hb.NH_2 \rightleftharpoons Hb.NH.COOH \text{ (carbohaemoglobin)}$$

Transport of CO_2 as Bicarbonates: The rest, or about 85 per cent of the total CO_2 is carried in the form of bicarbonates in both the plasma and red blood cells. As CO_2 enters the blood cells from the tissues, it combines with water to form carbonic acid (H_2CO_3), which dissociates to hydrogen ions (H^+) and bicarbonate ions (HCO_3^-). The latter diffuse into the plasma and with sodium or potassium ions in the plasma, form sodium or potassium bicarbonate.

$$Na^+ + HCO_3^- \longrightarrow NaHCO_3 \text{ (sodium bicarbonate)}$$

$$K^+ + HCO_3^- \longrightarrow KHCO_3 \text{ (potassium bicarbonate)}$$

It means at normal pH of mammalian blood, most of the total carbon dioxide will be present in the form of bicarbonate ion. According to Henderson-Hasselbalch equation, at pH 7.4, the ratio of carbonic acid to bicarbonate ion is 1: 20, *i.e.*, for each one part of carbon dioxide present as acid, twenty times as much carbon dioxide is present as bicarbonate ion. The hydrogen ion formed as a result of the ionic dissociation of the carbonic acid is buffered by the various buffering substances in the blood, and the effect of carbon dioxide on the pH of the blood is therefore only moderate.

Thus, CO_2 relies on the blood salts for most of its transportation. The red blood cells contain an enzyme, *carbonic anhydrase,* which increases the speed of reaction between CO_2 and H_2O, resulting in the formation of H_2CO_3 which is rapidly converted to bicarbonates as indicated above. Thus, the enzyme is responsible in large part for the fact that most of the CO_2 in the blood is ultimately converted to the form of bicarbonate ions. But like all other enzymes, carbonic anhydrase also catalyses the reversible reaction, that is the splitting of carbonic acid to water and carbon dioxide.

$$CO_2 + H_2O \underset{\text{(in lung capillaries)}}{\overset{\text{(in tissue capillaries)}}{\underset{\textit{carbonic anhydrase}}{\rightleftharpoons}}} H_2CO_3$$

Thus, the enzyme also accounts for the rapid release of CO_2 from the blood during its passage through the lungs. Thus despite the low concentration gradient (tissue Pco_2 is 45 mm Hg, that of arteriolar capillaries is 40 mm Hg) carbon dioxide diffuses rapidly from the tissues into the blood. This is a good example of facilitated diffusion.

Graph Representing Dissociation of Carbon Dioxide

By plotting on the ordinate the total amount of CO_2 in the blood and on the abscissa the various CO_2 partial pressures, a CO_2 dissociation curve may be constructed. However, the curves for oxygenated and deoxygenated blood differ somewhat. This is because oxyhaemoglobin is slightly more acidic than haemoglobin. Hence, oxygenated blood will bind slightly less CO_2.

This phenomenon is closely related to the *Bohr effect* which shows the other side of the coin, *i.e.*, an increase in the CO_2 shifts the oxygen dissociation curve to the right giving more oxygen from the blood to the tissues. When CO_2 is added to the blood, it pushes the equilibrium between oxyhaemoglobin and haemoglobin in the direction of the weaker acid (haemoglobin). This, therefore, tends to release more oxygen which, in fact, is the Bohr effect.

In the body the actual dissociation curve for CO_2 will be the fully drawn line A-V, in which the point A stands for the normal arterial blood and the point V for the mixed venous blood which is never completely deoxygenated and this curve therefore is said to be the *functional curve* or the physiological dissociation curve.

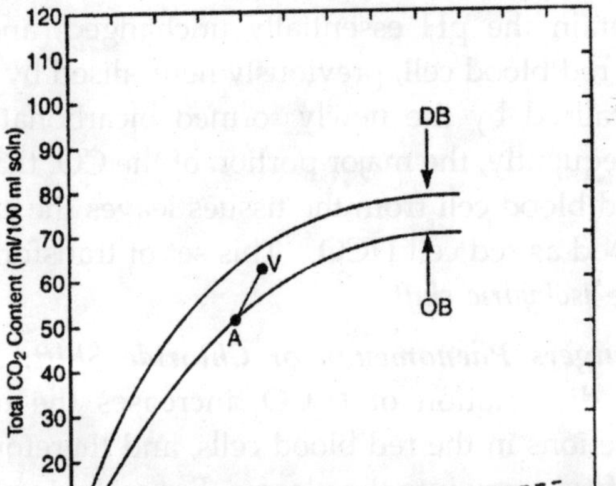

The carbon dioxide dissociation curve of mammalian blood (Winton and Bayliss, 1955). The dashed line indicates the amount of CO_2 held by physical solution in water. DB, curve for deoxygenated blood. OB, curve for oxygenated blood. A—V, physiological dissociation curve.

Isohydric Change

When arterial blood arrives in the tissue capillaries, about 95 per cent of the Hb is oxygenated. Because of the increased CO_2 tension and the decreased O_2 tension, the $HbCO_2$ dissociates; O_2 diffuses out into the interstitial fluid as CO_2 diffuses into the red blood cell. A small fraction of CO_2 is immediately bound as carbaminohaemoglobin. However, there still remains a large excess of CO_2 to be disposed of by other means. Under the influence of the enzyme *carbonic anhydrase,* this CO_2 is rapidly hydrated to H_2CO_3, which then dissociates.

Then two opposing phenomena take place: (1) the H_2CO_3 would tend to lower the pH within the red blood cell, but (2) the transformation of HbO_2 to Hb involves an apparent change of pK from 6.2 to 6.6, which tends to raise the pH within the red blood cell. As a result, protons formed in the dissociation of H_2CO_3 are accepted by the groups in Hb participating in the Bohr effect described above. The net result of these two events is to maintain the pH essentially unchanged, and K^+ ions within the red blood cell, previously neutralised by HbO_2, are now neutralised by the newly formed bicarbonate (HCO_3^-) ions. Consequently, the major portion of the CO_2 that diffused into the red blood cell from the tissues leaves the capillary in venous blood as red cell HCO_3^-. This set of transformations is termed the *isohydric shift.*

Hamburgers Phenomenon or Chloride Shift: As shown above the dissociation of H_2CO_3 increases the number of bicarbonate ions in the red blood cells, and therefore the ions tend to diffuse away into the plasma. For each bicarbonate ion that comes out from the red cells, one negatively charged chloride ion present in the plasma moves into the red cells in order to maintain acid-base equilibrium for the blood and the electrical neutrality of the red blood corpuscles.

Thus, *chloride shift involves the passage of chloride ions from the plasma into the red blood corpuscles to balance the bicarbonate ions that have passed from the red blood corpuscles into the plasma.* The carbon dioxide capacity of the red blood corpuscles is further increased by the chloride shift, because the removal of bicarbonates, in this way, enhances their formation from the carbonic acid.

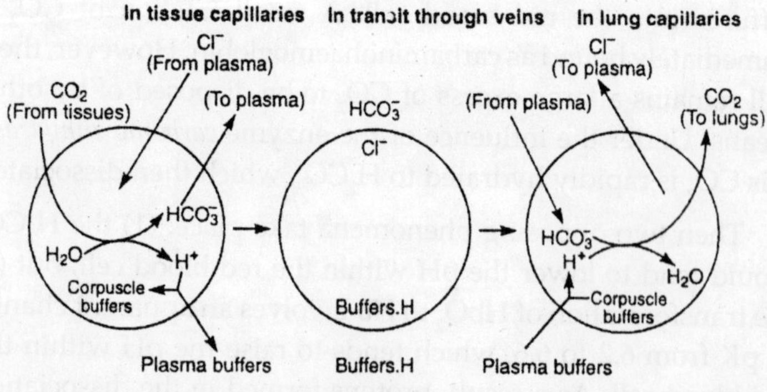

The Chloride shift.

In the red blood corpuscles most of the buffering is provided by the haemoglobin (Hb) itself. The latter is a negatively charged blood protein and combines with positively charged hydrogen ions (formed in the ionisation of (H_2CO_3), forming the haemoglobinic acid. This reaction is shown as follows:

$$BHb + H^+ \rightleftharpoons \underset{\text{(haemoglobinic acid)}}{HHb} + B^+$$

The resulting cation (B^+) from haemoglobin is balanced by the Cl^-.

Formation of CO_2 and its Release in the Body

The CO_2 is carried by the blood-stream to the lungs in the form of carbonic acid, bicarbonates and carbamino compounds. Due to a higher tension of CO_2 in the lung capillaries (than the tension of CO_2 in the lung alveoli), CO_2 is given off from the blood and the reactions taking place are as follows:

(i) $H_2CO_3 \xrightarrow{\text{carbonic anhydrase}} H_2O+CO_2$

(ii) $NaHCO_3 \xrightarrow{\text{acidity of oxyhaemoglobin}} Na^+ + HCO_3^-$

or $2NaHCO_3 \xrightarrow{\text{acidity of oxyhaemoglobin}} Na_2CO_3 + H_2O + CO_2$

(iii) $HbNHCOOH \xrightarrow{\text{increasing oxyhaemoglobin}} HbNH_2 + CO_2$

The reciprocal effect of oxygenation on acid strength of the haemoglobin, the so-called *Haldane effect,* accounts for the CO_2 exchange. In the lungs, chloride ions move out of the red blood corpuscles and bicarbonate ions move back in. The enzyme carbonic anhydrase then promotes the rapid reformation of free CO_2, and this gas diffuses from the blood into the lung alveoli.

Cellular (internal) Respiration and Oxidation: The ultimate aim of all the respiratory activities is the cellular oxidation, *i.e.,* the aerobic breakdown of the digested food materials and the release of energy thereby. The breakdown of carbohydrate may be represented by the following overall equation:

$$C_6H_{12}O_6 + 6O_2 \longrightarrow 6CO_2 + 6H_2O + \text{energy (686 Kcal/mole)}$$

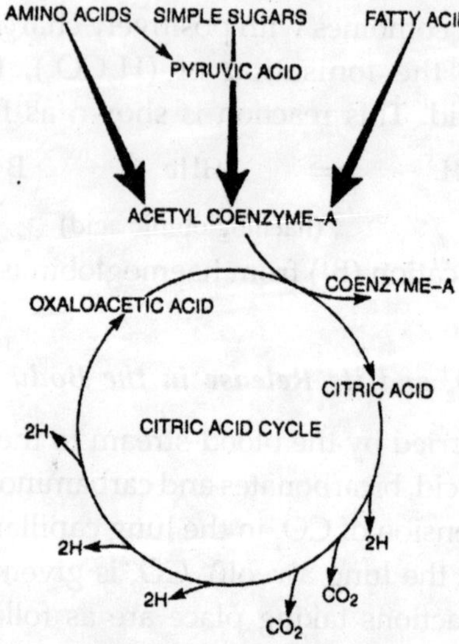

The interrelationship of metabolic pathways with Kreb's cycle.

The interrelationship of metabolic pathways of amino acids, simple sugars and fatty acids with the Kreb's cycle is summarised in Fig. Thus, within each cell glucose is first metabolised to pyruvic acid and then, if there is supply of O_2 to CO_2 and H_2O. It is when hydrogen atoms are combined with O_2 that most of the energy is released in the cell. By storing energy as ATP molecules, the cell can utilise this energy for its various activities as needed.

Nervous and Chemical Control of Respiration: Both the rate and depth of respiration can be regulated by chemical and neural factors. The process of breathing is essentially involuntary and automatic. However, it is influenced and controlled voluntarily within limits. Normally inspiration and expiration follow each other and both are basically muscular in action. The contraction and relaxation of the respiratory muscles (diaphragm and several chest muscles) depend on the arrival of nerve impulses originating from a group of neurons collectively forming the *respiratory centre* located in the medulla oblongata of the brain. The respiratory centre consists of the following parts.

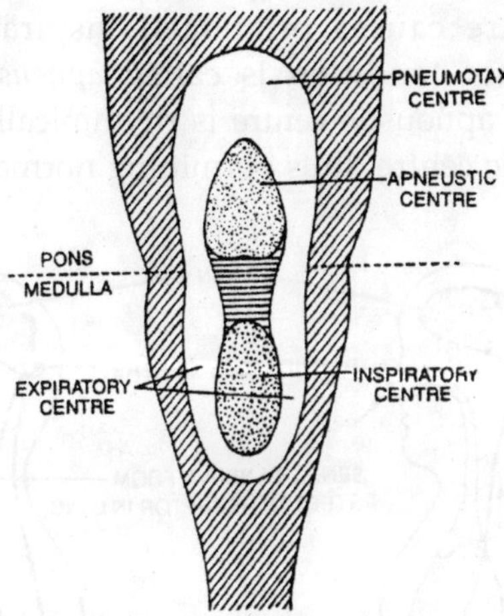

The respiratory centre.

Inspiratory Centre: The inspiratory neurons are grouped to form an inspiratory centre in the medulla oblongata. Motor neurons from this centre innervate the muscles of diaphragm and external intercostal muscles. The inspiratory centre probably originates the respiratory impulses which are modified by higher centres and also by impulses from spinal afferent nerves, specially those from the lungs.

Expiratory Centre: The expiratory neurons are grouped to form an expiratory centre in the medulla oblongata. Motor nerves from this centre innervate the abdominal muscles and internal intercostal muscles.

Pneumotaxic Centre: Located in the upper pons is a component of an oscillating circuit between the inspiratory and expiratory centre and itself. This centre does not originate impulses, but is stimulated by the activity of the inspiratory centre. When it is active, it stimulates the expiratory centre in the medulla oblongata and inhibits the other pontine centre (the apneustic centre). This facilitates expiration.

Apneustic Centre: The apneustic centre in the middle and lower pons stimulates the inspiratory centre in the medulla and causes inspiration. Prolonged electrical stimulation of the apneustic centre causes prolonged inspiration (without expiration). This condition is called *apneusis.* In normal respiration, the apneustic centre is rhythmically inhibited by the pneumotaxic centre, thus permitting normal expiration to occur.

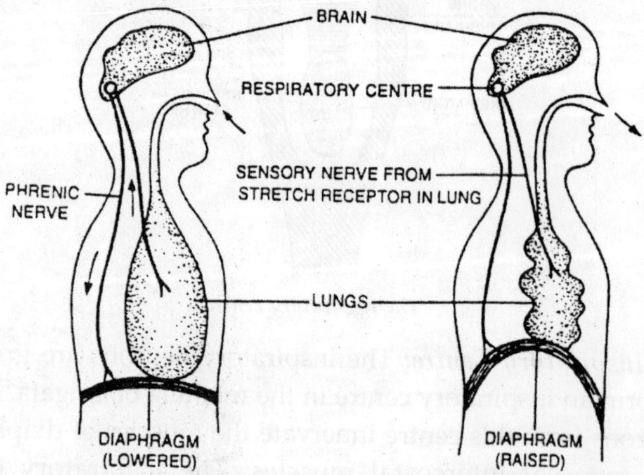

The control of respiration by respiratory centre. (A) CO_2 in blood stimulates the latter to send impulses to the diaphragm, leading to inhalation. (B) Impulses from the inflated lungs inhibit the breathing centre, leading to exhalation.

The reflex control of breathing depends mainly on the following reflexes (mechanisms):

Hering-Breuer Reflex: During inspiration, the inspiratory centre sends out nerve impulses which cause the contraction of the diaphragm and other respiratory muscles. As a result the thorax and lungs expand. Then some stretch-sensitive receptors (Hering-Breuer receptors) in the lung tissue are stimulated as soon as the expanding lungs attain a certain threshold tension. The nerve impulses thus initiated at the stretch-sensitive receptors in the lung are propagated by afferent neurons in the vagal nerves to the inspiratory centre which, as a result, is inhibited. The impulses may also go directly to the expiratory centre which responds to them by giving impulses that inhibit the inspiratory centre. The latter, therefore, stops discharging impulses to the concerned muscles which consequently relax causing expiration. This mechanism is called the Hering-Breuer reflex, first described by them in 1868.

Pneumotaxic Reflex: The pneumotaxic or pontine centre is believed to function at the instigation of the inspiratory centre by sending impulses to the expiratory centre in almost the same manner as do the stretch receptors in the lungs. The stimulated expiratory centre in turn inhibits the activity of the inspiratory centre as indicated above. This mechanism is called the Pontine mechanism or the *pneumotaxic reflex.*

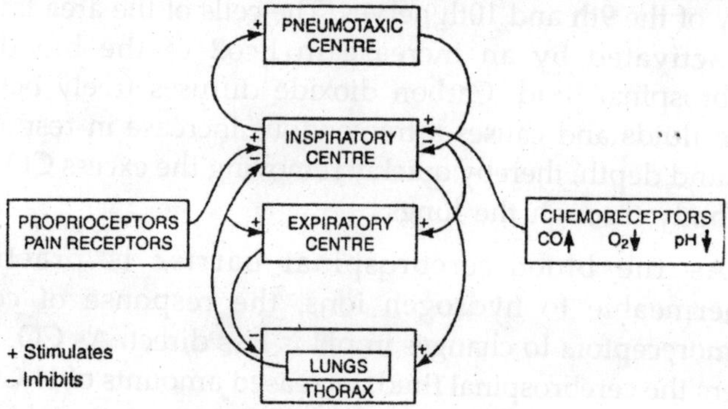

Reflex control of breathing. Reflexes are initiated by the proprioceptors, chemoreceptors, pain receptors, stretch of the lungs, and action of the inspiratory centre itself via the pneumotaxic centre or the expiratory centre.

Carotid and Aortic Body Reflexes (Chemoreceptor Respiratory Reflexes): *Heymans* (1936) demonstrated that *carotid bodies* (small vascular spherules located at the bifurcation of the external and internal carotids) and similar structures on the aorta, *aortic bodies,* were chemosensitive. Present evidence suggests that the chemosensitive cells of these bodies are active at the normal Po_2 of the arterial blood. However, a great increase in firing rate occurs when Po_2 is reduced to 50 mm Hg (normal blood Po_2 in the arteries is about 100 mm Hg). This fall in Po_2

increases the frequency and amplitude of nerve impulses from the carotid body nerves to the respiratory centres in the medulla. The resulting increase in alveolar respiration is a feedback mechanism that raises arterial Po_2. Thus, *peripheral chemoreceptors* (carotid bodies and aortic bodies) respond to changes in Po_2. The Pco_2 plays a role in the sensitivity of these cells to Po_2, because when Pco_2 is increased, the response to decreasing Po_2 is exaggerated. Thus, asphyxia, which reduces Po_2 and increases Pco_2 is a circumstance in which the receptors are greatly activated. Hypotension (low blood pressure) is another circumstance that may elicit chemoreceptor activity.

Central Chemoreceptors: The central chemoreceptor areas *(area lateralis)* of the medulla are found on either side near the entry of the 9th and 10th nerves. The cells of the area lateralis are activated by an increase in Pco2 of the blood and cerebrospinal fluid. Carbon dioxide diffuses freely between these fluids and causes a tremendous increase in respiratory rate and depth, thereby quickly removing the excess CO_2 from the body through the lungs.

As the blood-cerebrospinal barrier is practically impermeable to hydrogen ions, the response of central chemoreceptors to changes in pH is less direct. As CO_2 freely enters the cerebrospinal fluid, increased amounts of H_2CO_3 are formed. Consequently, as this acid dissociates, the pH of the cerebrospinal fluid is increased which serves as a stimulus for increased activity of the central chemoreceptors.

Effect of Exercise on Respiration

There is an increase in the rate and depth of breathing during physical exercise. In moderate exercise, ventilation increases in direct proportion to the amount of CO_2 formed. But during strenuous exercise, lactic acid is also formed, and the resulting increase in arterial pH acts as an additional stimulus to pulmonary ventilation. These changes in Po_2, Pco_2 and pH *(humoral factors)*, however, can account for only a small amount of increased respiration during exercise.

By ingenious cross-circulation techniques exercise results in a marked increase in respiration even when the Po_2, Pco_2 and pH are artificially maintained at resting levels. This increase is caused by impulses reaching the respiratory centre from *proprioceptors* in the active skeletal muscles. Passive "pumping" of these limbs to excite these proprioceptors also increases pulmonary ventilation.

A third source of increased respiration during exercise is the *motor cortex,* which responds to the stress of exercise by exciting the respiratory centre by way of the reticular formation. This is similar to the role of the cortex in increasing cardiac output during exercise. As a result, the respiratory and circulatory adjustments to exercise are well integrated through these higher levels of the nervous system.

Other Factors Affecting the Respiratory Centre: The respiratory centre is also influenced by some chemical and physical factors. The partial pressure of CO_2 in the blood is probably the most important, for the respiratory centre is highly sensitive to it. A little rise in the CO_2 tension stimulates the respiratory centre, resulting in an increased rate of breathing till the tension comes to normal. Increased acidity of the blood also has the same stimulating effect on the respiratory centre as does the CO_2.

The increased concentrations of CO_2 and lactic acid during muscular exercise are responsible for the accompanying increase in the rate and depth of breathing. The breathing rate is also influenced by the body temperature and blood pressure. A rise in temperature stimulates breathing and vice versa. A decrease in blood pressure increases the breathing rate, whereas an increase causes a reverse effect.

In case of the voluntary control of breathing, impulses are transmitted from the higher centres of the brain (cerebral cortex) to the respiratory centre. Similarly, strong emotions, fright, excitement and other mental factors also have a striking influence on the breathing process.

Effect of Height on Respiration

At high altitudes the oxygen pressure may be quite low, 60 instead of 160 mm Hg. Consequently, there is a reduction in P_i, O_2 and symptoms of *hypoxia* (lack of O_2) appear. These symptoms include nausea, dizziness, loss of memory, cyanosis (blueness of lips and nails), and changes in heart rate and pulmonary ventilation. These symptoms are seen in varying degrees of severity in *mountain sickness*, depending on the altitude and the extent of acclimatisation. These symptoms are exacerbated by exercise, which increases the tissue demand for oxygen. Therefore, it is usually necessary to breathe oxygen at 6500 m. In order to maintain the haemostatic mechanisms between respiration and blood, a mixture of O_2 (95 per cent) and CO_2 (5 per cent) is usually administered.

In respiration at high altitudes ventilation increases, which substantially reduces the difference between inspired Po2 and Po2 in the alveolar gas. Associated with this rise in ventilation there is also a reduction in Pa, CO2 which gives rise to *alkalosis* (fall in the alkali reserve of the blood) in new arrivals to a high attitude. Over a period by few days, the kidneys remove HCO_3- and blood pH is returned to normal. The diffusion capacity of the lung increases such that P_A, O_2-Pa, O_2 is less in residents at high altitude than sea-level residents. This is at least partly related to an increase in the ratio of alveolar volume to total lung volume, and may also be the result of a reduction in the blood-air diffusion distance.

Adaptations also occur in the circulatory system. Both cardiac output and O_2 carrying capacity of the blood are higher in animals living at high altitude. The latter is related to a rise in haematocrit. The O_2-affinity of the blood of a number of mammals at high attitude, including human, decreases, and this is related to an increase in the level of 2, 3-diphosphoglycerate (DPG). This helps in the removal of O_2 from the blood at the tissues. The blood of vicuna and llama has a higher O_2 affinity than that of mammals living at sea level, so the response of the dissociation curve to hypoxia varies from animal to animal (*Lenfant*, 1973). At high altitudes, the number of red cells in the blood is increased, a condition termed *polycythemia*.

Definition of Respiratory Quotient

The ratio of the volume CO_2 given off to the volume of O_2 consumed during the same time is called the respiratory quotient (R.Q.). The R.Q. can be calculated for the oxidation of various food stuffs. In the reaction given below, *i.e.*, for carbohydrate oxidation it is 1.00.

$$C_6H_{12}O_6 + 6O_2 \longrightarrow 6CO_2 + 6H_2O + \text{energy}$$

$$RQ = 6 \text{ volumes } CO_2/6 \text{ volumes } O_2 = 1.00$$

It varies with different foods, for fats the R.Q. is about 0.7; for proteins about 0.8, and for the average mixed diet about 0.85. It was formerly thought that the R.Q. of an organism or tissue gave direct information concerning the kind of food stuff being oxidised, or, in other words, of the kind of metabolism that is taking place in the body. But in practice the interpretation of *R.Q.* is not so simple, for cells can oxidise more than one of the three food stuffs at the same time, and certain other metabolic processes may use O_2 or produce CO_2.

Role of Respiratory Pigments

These are the substances which combine reversibly with O_2, thus acting as a carrier or store of it. In this way, whenever a circulatory system transports O_2 from a respiratory surface to the body tissues, there is usually a transport pigment in the blood. One factor regarding the mode of functioning of these pigments is their O_2 affinity, which is indicated by the partial pressure of O_2 at which they are half-saturated with O_2. If the value (referred to as the p50) is high, the pigment has a low affinity, for it will give up its O_2 while this is still relatively plentiful in the surrounding medium. If the *p50* is low, the pigment has a high affinity, for it will only yield up its O_2 when the partial pressure of O_2 in the surrounding medium has become very low. All respiratory pigments contain a metal in an organic complex. A brief account of different respiratory pigments met within the animals is given below:

Haemoglobin: Haemoglobin is the best known O_2-carrying pigment. It occurs in the plasma of some animals and in the red blood cells of others. For example, certain annelids, molluscs and crustaceans carry haemoglobin in the plasma. A few annelids are known in which the pigment occurs both in plasma and blood cells. Other annelids, nemertine worms, sea cucumbers and vertebrates are among animals in which haemoglobin is found in the blood cells. With some very rare exceptions, *e.g.*, the *leptocephalus* (larva of the eel), all classes of vertebrates possess haemoglobin in their red blood cells and their muscles contain *myoglobin* (muscle haemoglobin or *Mb*).

Haemoglobin molecules from wide variety of species differ qualitatively in such properties as solubility, crystal form, affinity for O_2, and absorption spectra. But all are composed of a colourless protein *globin* combined with *haeme.* Such differences in properties are due entirely to the different amino acid sequences and the conformations of the globin ; the haeme moiety is identical in all vertebrate and most invertebrate haemoglobins.

Haemoglobin is a conjugated protein *(chromoprotein).* Each erythrocyte contains about 100 million haemoglobin molecules. Each unit of haemoglobin is actually a combination of an iron-containing *protoporphyrin (Heme)* and a specific protein globin. The unit molecular weight is approximately 17 to 18 $\times$ 10^3. The blood haemoglobin of higher vertebrates has four units *(i.e.,* molecular weight = 68 to 72×10^3).

The protein moiety or globin varies considerably in size, amino acid composition, solubility, and other physical properties from animal to animal. The iron content of mammalian haemoglobin is 0.336 per cent and the heme content is 4 per cent, the iron being normally in the ferrous state. In each unit of heme, the iron atom is joined by four of its coordination bonds to the four nitrogen atoms of the protoporphyrin and one of the remaining six coordination bonds is joined to a molecule of globin. Thus, the chemical structure of haemoglobin consists of four molecules (or units) of heme combined with the globin molecule.

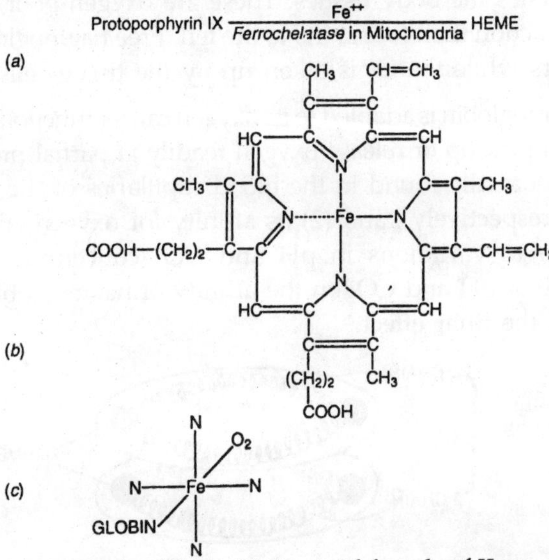

(a) Synthesis of Heme, (b) Structural formula of Heme,
(c) Linkage of ferrous iron, oxygen, and globin in a unit of haemoglobin.

Haemoglobin uncombined with O_2 and containing haeme with ferrous iron (Fe^{2+}) is called *ferrohaemoglobin, deoxyhaemoglobin,* or *reduced haemoglobin* and is abbreviated Hb. Human adult haemoglobin (HbA) or normal haemoglobin is composed of four polypeptide chains, two α-chains and two β-chains, and its subunit structure is designated $\alpha_2 A \beta_2 A$. Each unit can carry an O_2 molecule attached to the remaining coordination bond. Since each of the four hemes contains an iron that serves as the point of loose physical attachment for a single molecule of O_2, a mammalian haemoglobin can carry a maximum of four molecules of molecular oxygen. When haemoglobin (Hb) is fully oxygenated, it is called *oxyhaemoglobin*

(HbO_2) which contains 4 molecules of O_2 per molecule of haemoglobin.

$$Hb_4 + 4O_2 \; 5{=}\rightleftharpoons Hb4(O_2)_4 \text{ (oxyhaemoglobin)}$$

This reaction proceeds to the right when O_2 is present in excess, as in the lungs. Oxyhaemoglobin is carried in blood and thus reaches the body tissues. These are oxygen-poor, and the above reaction therefore shifts to the left. Free haemoglobin then reappears while the O_2 is taken up by the tissue cells.

Haemoglobin is adapted to its oxygen carrier function because: (1) it can pick up or release oxygen readily at partial pressure of oxygen normally found in the blood capillaries of the lungs or tissues, respectively ; and (2) its affinity for oxygen changes in response to variations in pH and CO_2 concentrations. The influence of pH and CO_2 on the affinity of haemoglobin for O_2 is called the Bohr effect.

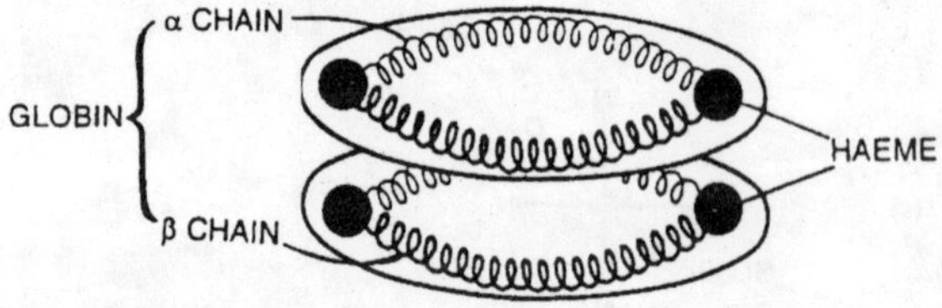

The structure of the haemoglobin molecule formed by four polypeptide chains and four molecules of haeme distributed to two indentical subunits.

Haemoglobin also functions as a carrier of small quantity of CO_2, conveying it as a waste product from the tissues to the lungs for excretion. However, the CO_2 is not transported by the haemoglobin's iron, but instead is carried in part in combination with the protein of the haemoglobin molecule.

Haemoglobin can also combine with 4 molecules of carbon monoxide (CO) to give carbon monoxide haemoglobin or *carboxyhaemoglobin* (HbCO), which is photosensitive and dissociates in light to release CO. The affinity of human Hb for CO is more than 200 times greater than its affinity for O_2. The brilliant cherry-red colour of HbCO is quite distinctive and is manifested in the skin and tissues of victims of CO poisoning.

Four molecules of NO also combine with one molecule of Hb to give the very stable HbNO. Many agents, *e.g.*, peroxides, ferricyanide, can oxidise the Fe^{2+} in Hb to Fe^{3+} to give *methaemoglobin* (met Hb), which binds neither O_2 nor CO. *Cyanomethaemoglobin* is formed on reaction of met Hb with CN^-.

When haemoglobin content of the blood is reduced, the result is anaemia. Haemoglobin is not only a carrier of O_2 and CO_2, but it is also one of the buffering agents in the blood. Some abnormal haemoglobins (haemoglobins, with altered solubility, altered O_2 affinity or unstable haemoglobins) also occur. For example, individuals homozygous for HbS, abnormal haemoglobin (sickle cell anaemia) have the severe disease *sickle-cell anaemia* and those heterozygous for HbS and HbA (normal haemoglobin) have *sickle-cell trait*, a condition usually without clinical symptoms.

Myoglobin: Myoglobin (Mb) of vertebrate muscle is very similar to Hb. While haemoglobin is an *oxygen carrier* protein, myoglobin is an *oxygen storage* protein. It stores O_2 in resting skeletal muscle as *oxymyoglobin* (MbO_2) and on exercise releases O_2. Myoglobins have only about 30 residues homologous with either the a or b chains of vertebrate haemoglobins, but their tertiary structures are strikingly like those of the Hb chains, with the same kinds of helical and non--helical secondary structures. The most important structural differences between Hb and Mb, however, are the distributions of R groups on the surfaces of the molecules. In haemoglobin, there are many surface R groups that provide complementary contacts between subunits and allow association of the chains to form the native quaternary structure, a_2Ab_2A. Since the R groups that provide these contacts in a and b chains are absent in Mb, in dilute solution Mb is a monomer and Mb cannot replace an a or b subunit in Hb.

In muscle, myoglobin binds O_2 reversibly. However, it exhibits no Bohr effect (which is exhibited by Hb) and its O_2 dissociation curve is hyperbolic (which is sigmoidal for human and frog haemoglobin). During muscle contraction, when the demand for O_2 is highest, O_2 dissociates from myoglobin and is available for oxidations. In humans, myoglobin is abundant only in cardiac muscle. In diving mammals, the myoglobin content of muscles is very high and facilitates submersion for long periods. The flight muscles of birds are also rich in myoglobin. At a venous Po_2 of 40 mm Hg, at which Hb is 66 per cent saturated, myoglobin is still 94 per cent saturated. At an O_2 tension of 10 mm Hg, Hb is 10 per cent saturated whereas myoglobin is 80 per cent saturated. Therefore, cytochrome oxidase can operate at Vmax when the medium provides O_2 at Po_2 of about 4 to 5 mm Hg.

Thus, the affinities of these three proteins for O_2 are in the order cytochrome oxidase > myoglobin > Hb. As a result, myoglobin can accept O_2 from Hb and store it in the muscle cell for release to cytochrome oxidase when the O_2 supply is limited.

Erythrocruorin: Large haemoglobins, frequently called erythrocruorins, occur in the circulating plasma of many annelid worms and molluscs. The giant molecule of erythrocruorin from *Limnodrilus*, consists of 108 subunits (MW = 2.8×10^4), each bearing one haeme, arranged as a regular hexagonal cylinder having the dimensions 220 by 160 Å. This cylinder, in turn, consists of six lesser cylinders oriented in the same axis as the major cylinder axis. Thus, there are three layers of 36 protein subunits each in the overall packaged molecule. Since there are four different NH_2-terminal amino acids, it is assumed that there are four different subunit polypeptide chains in the molecule.

Chlorocruorin: Certain of the annelids, such as *Spirographis*, have green blood pigment, the chlorocruorin. In this pigment, the porphyrin differs from protoporphyrin in that the 2-vinyl group is replaced by a formyl group. When such respiratory pigments are present in the coelomic fluid rather than in an organised circulation, they serve to store rather than transport O_2.

Haemerythrin: Haemerythrin is present in the blood of all sipunculid worms, a few polychaete forms, and the brachiopod *Lingula*. It is a rare, reddish-violet, iron containing respiratory pigment. The most carefully studied haemerythrin from *Sipunculus* has a molecular weight of 105,000 and contains 16 atoms of Fe^{2+}. Haemerythrin binds one O_2 per pair of iron atoms.

Some Characteristics of Respiratory Pigments

S.No.	*Pigment*	*Source*	*Location*	*Atoms of metal per mole*	*Molecular weight*
1.	Haemoglobin	Human being	Intracellular	4Fe	6.5×10^4
		Frog	Intracellular	4Fe	6.5×10^4
		Gastrophilus larvae	Tracheal cells	2Fe	3.4×10^4
2.	Myoglobin	Mammalian muscle	Intracellular	1 Fe	1.7×10^4
3.	Erythrocruorin	Annelids	Plasma	96 Fe	3×10^6
		Molluscs	Plasma	96 Fe*	1.5×10^6
4.	Chlorocruorin	Polychaete worms	Plasma	96 Fe	3×10^6
5.	Haemerythrin	Sipunculus	Intracellular	16 Fe	10.5×10^4
6.	Haemocyanin	Squid	Plasma	20 Cu*	3.75×10^6
		Snail	Plasma	200 Cu*	6.76×10^6

* approximately

Haemocyanin: The blood plasma of many molluscs and arthropods contains *haemocyanin,* blue pigment containing copper (instead of iron), but no haeme. The molecular weights of haemocyanins range from 0.5 ¥ 10^6 to 10^7. Squid haemocyanin is a decamer (MW = 3.75 ¥ 10^6, which on dilution or at high salt concentration dissociates into 5 dimers and then into 10 monomers. Each monomer contains 2 atoms of Cu^+ and, both in monomeric and polymeric form, binds one O_2 per two Cu^+. There is some evidence that O_2 binding and release involve oxidation and reduction of the copper.

Respiratory pigments change colour according to the degree of oxygenation *(e.g.,* haemoglobin is red when oxygenated, and purple when deoxygenated) and many have characteristic absorption spectra. These properties have facilitated their investigation.

Significance of Anaerobic Respiration

When respiration goes on in the absence of O_2, it is called anaerobic. Fermentation in yeast and glycolysis in tissues and may microbes are typical examples of anaerobic respiration. The overall equation of fermentation can be written as follows:

$$C_6H_{12}O_6 \longrightarrow 2CH_3CH_2\ OH + 2CO_2 + \text{Energy (50 kcal/mole)}$$

The anaerobic breakdown of glucose in animal tissues differs from this in two respects: (i) it begins with glycogen stored in the tissues, for which reason it is called glycolysis, and (ii) it ends with lactic acid. The intermediate stages, however, are identical. The overall equation for glycolysis can be written as follows:

$$C_6H_{12}O_6 \longrightarrow 2CH_3\ CH(OH)\ COOH + \text{energy (36 kcal/mole)}$$

Such processes do not liberate as much energy as does aerobic respiration. Many organisms or parts of them respire anaerobically for sometime when their supply of O_2 is insufficient for aerobic respiration, in which case they may acquire an oxygen debt. A few bacteria are strictly anaerobic, *i.e.,* they do not use free O_2 at all.

Oxygen Consumption and Metabolic Rate Determination

Basal Metabolic Rate: The term basal metabolic rate (BMR) refers not to the lowest possible energy metabolism of an organism but to a *minimum stable rate of metabolism* when the animal is in the resting condition. In birds and mammals, BMR is measured when they are in the resting and fasting state so that the digestive processes are not going on which would require the expenditure of energy. This rate can be determined in three ways:

(a) From the food consumption or by determination of the difference between the energy value of all the food taken in and the energy value of faeces and urine.

(b) From the energy released as heat, or from the total heat production of the organism.

(c) From the amount of oxygen used in oxidation process to obtain energy, provided the information is available about which substances have been oxidised and there is no anaerobic metabolism or external work performed.

But all the three methods are not equally satisfactory. However, the third method is technically easy and is widely used for this purpose. So when we say metabolic rate, we actually mean the rate of O_2 consumption. A simple apparatus to determine O_2 consumption is given in Fig.

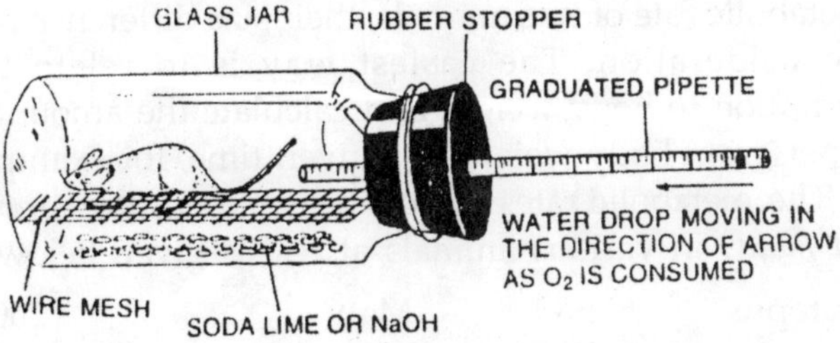

Simple apparatus to measure the rate of oxygen consumption.

As O_2 is used by the animal, the volume of the air in the jar decreases, for the CO_2 is absorbed by alkali as fast as it is formed. The movement of the drop in the graduated pipette shows the reduction in volume, and by timing its movement we get the number of ml O_2 used by the animal per minute. Further calculations can be made as shown below:

Weight of the apparatus without animal = *a* gm.

Weight of the apparatus including animal = *b* gm.

Weight of the animal *(b-a)* = *c* gm.

Time taken for the coloured drop of water to travel from 0.1 ml mark to 1.0 ml mark= *t* sec.

Thus, 0.9 ml of O_2 is consumed in *t* seconds by c gm of animal tissue (at room temperature A degree centigrade and B mm Hg pressure).

Therefore, O_2 consumption per hour by 1 gm of animal tissue (at A° centigrade and B mm Hg pressure).

$$= \frac{0.9 \times 60 \times 60}{t \times c} = \frac{3240}{tc} \text{ ml}$$

Standard Metabolic Growth

Poikilothermic animals have metabolic rates that are temperature dependent and therefore have no metabolic state corresponding with the avian or mammalian BMR. So for poikilothermic animals *the minimum fasting metabolism at a given temperature* is called the standard metabolic rate (SMR).

Metabolic Rate in Various Animals: For a comparison of the metabolic rate of two animals, their size difference is taken into consideration. The easiest way is to relate the O_2 consumption to body weight and calculate the amount of O_2 used per gram body weight in a given time, for example one hour. The metabolic rates *(O_2 consumption in* mm^3 */gram body weight /hour)* in various animals at rest is given below:

Octopus	80	Man	200
Squid	320	Rat	1500
Eel	128	Humming bird (at rest)	1000
Trout	226		
Frog	150	Humming bird (On flight)	13,000

The metabolic rate is calculated when the animal is in the resting condition, for any activity requires extra energy and the metabolic rate is naturally increased. A comparison of the metabolic rate of the Humming bird at rest and on flight makes this point clear. This is the reason that active animals always have high metabolic rates and therefore, some of the highest metabolic rates on record have been found in the flying insects and birds.

The metabolic rate is also related to body size as small animals have a high metabolic rate. For example, in a rat of 25 grams the oxygen consumption in mm^3/gram/hour is about 1580, in a rabbit of 2200 grams, 466, and in a man weighing 70,000 grams it further decreases to only 202. Thus, it is clear from these data that the metabolic rate increases with the decrease in body size (or weight). However, there should be certain limit of size within which this increase is possible.

The smallest existing mammals are some shrews, having a body weight of only 4 grams, but their metabolic rate is very high. To maintain this rate, a shrew has to eat constantly and is always on a search for food, consuming almost its own body weight of food per day. If a mammal were even smaller, it would have to eat at such a tremendous rate and spend so much energy in search of food that it simply could not exist. It can therefore, be concluded that there is a lower theoretical limit to the size of mammals.

However, since metabolic rate actually decreases with the increasing body size, the limit to the body size of mammals is not set by the requirements for food. Land animals cannot increase to unlimited size because the mechanical strength of their muscles and bones could not support their enormous weight. The heavy bones and thick legs of elephant are evidence of the strength needed to support a huge body, and a further increase in weight is just not possible without the animal becoming overly cumbersome.

Metabolic Rate and Circadian Rhythms: One of the most remarkable attributes of organisms is the rhythmic or cyclic nature of most of their functions and activities. Many of these endogenous rhythms have a periodicity of 24 hrs. In an evolutionary sense this 24 hour periodicity is undoubtedly related to the period of the earth's rotation and hence the length of one light-dark cycle. In animals under natural conditions such a 24 hour periodicity is daily entrained or reinforced by the normal day-night cycle.

However, the absence of daily reinforcement does not lead to the loss or disappearance of biological rhythmicity. Majority of animals maintain a 24 hour periodicity even when kept in an unchanging environment in which they receive no clues about the natural day-night cycle. One of the most striking features of these rhythms is their insensitivity to environmental temperature. Even in poikilotherms wide variations in environmental temperature have only slight effect on the daily periodicity of behaviour or function.

In the absence of an external timegiver *(zeitgeber)* these endogenous rhythms rarely have a period that is exactly 24 hours. Instead, it varies between 22 and 26 hrs. Any given individual animal, however, keeps a constant periodicity to its daily rhythm that may be slightly shorter or slightly longer than the environmental cycle. *These daily endogenous cycles are called circadian rhythms (L. circum* = about, *dies* = *day) because they are about one day (though not exactly) in length.*

It is known that nearly any measurable physiological quantity, *e.g.*, body temperature, oxygen consumption, blood sugar, *etc.*, shows a striking daily cycle. Because all these and many other factors vary in a cyclic manner, time of day must be taken into account in virtually all physiological measurements. Oxygen consumption, activity, and body temperature can be monitored over long periods of time without disturbance, so the effects of circadian rhythmicity are well documented with regard to these functions.

Typically the body temperature of diurnal homeotherms is 1 to 3° higher during the day than at night. For nocturnal forms the cycle is reversed. The daily rhythms of the metabolic responses of vertebrates are complex than those of body temperature but they generally show the same fundamental pattern, with rates being high during the normal period of activity and low during the normal period of inactivity. The persistence of such endogenous rhythms allows the animal to be ready to function adequately before the environmental stimulus appears, *i.e.*, it is anticipatory and thus enhances the effectiveness of responses *(Aschoff*, 1960).

Oxidations in Biology

Characteristics and Examples

(1) In contrast to explosive, inefficient, and uncontrolled oxidation in ordinary combustion of fuels, biological oxidation is stepwise, efficient and controlled, as in the latter case, energy is released gradually by breaking

the chemical bonds in the food molecule one after another. A respiratory decomposition does not occur in a single large oxidation-reduction process, as just shown for methane. Instead, respiration takes place through a series of consecutive redox reactions in which H is removed from fuel in several steps, each resulting in a successively higher oxidation level of carbon atoms. Such stepwise oxidation occurs generally in the respiration of all foods. The carbon groupings of food molecules are transformed successively from hydrocarbon-to-alcohol-to aldehyde-to acid-to anhydride. Such stepwise oxidations provide energy in smaller successive packs and each of these can become useful in the cellular metabolism.

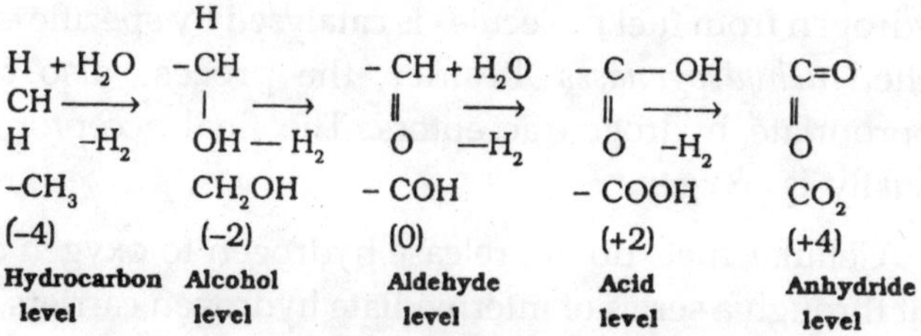

Foods and Oxidation Levels

Besides, the oxidation states of carbon here change successively from -4 to -2 to 0 to +2 to +4.

(2) Most of the energy of combustion dissipates as heat, v whereas a large fraction of the energy released in biological oxidation is captured in ATP. A very small amount of energy which is not captured, is liberated as heat. This too, need not be considered as "wasted", since in the warm-blooded animal it contributes to the maintenance of body temperature.

The Biological oxidation may take place in the following three ways:

(i) *By the addition of* O_2

$$C + O_2 \rightarrow CO_2$$

(ii) *By the removal of hydrogen*

$$\underset{\text{Lactic acid}}{CH_3CHOH.COOH} \rightarrow \underset{\text{Pyruvic add}}{CH_3CO.COOH} + 2H^+ + 2e^-$$

(iii) *By an increase in positive valency*

$$\underset{\text{Ferrous chloride}}{2Fe^{2+}(Cl-)_2} + Cl_2 \rightarrow \underset{\text{Ferric chloride}}{2Fe^{3+}(Cl^-)_3}$$

Biological oxidation is often synonymous with dehydrogenation, i.e., removal of hydrogen atoms and electrons.

Hydrogen Transfer (Electron Transport System): There are two kinds of ATP sources in respiration. One is nutrient oxidation by dehydrogenation, the other is creation of ATP during hydrogen transfer from fuel to oxygen. Removal of hydrogen from fuel molecules is catalysed by specific enzymes called *dehydrogenases*. Besides, the process also requires appropriate hydrogen acceptors. The final acceptor in cells usually is oxygen.

Cellular fuels do not release hydrogen to oxygen directly, but through a series of intermediate hydrogen carriers, the last of which finally releases hydrogen to oxygen. The advantage of such step by step transfers is that more usable energy may be obtained for metabolic use, than if hydrogen and oxygen were allowed to combine directly. The combination of hydrogen and oxygen is an energy-yielding process. Thus for every H_2 transferred by carriers from fuel to oxygen, a net total of 3 ATP moles is formed.

The sequence of enzymes and carriers responsible for transport of reducing equivalents (H or electrons) from substrates to molecular oxygen is known as the *respiratory chain* or *Electron transport System* (ETS). The respiratory chain is now known to be localised within mitochondria. The major components of the respiratory chain are arranged sequentially in order of increasing redox potential. Electrons flow through the chain in a stepwise manner from the more electronegative components to the more electropositive oxygen.

Every step in serial hydrogen transfer is itself a complete redox process. The intermediate hydrogen carriers are alternately reduced and oxidised, *i. e.*, they operate in a cyclic manner, first accepting H (or electrons), then passing it on, and then again accepting new hydrogen, etc. Hydrogen transfer by carriers yields energy at each transfer step, but at three of these steps, enough energy is liberated to generate ATP.

Electrons are transferred from NADH or $FADH_2$ to molecular oxygen (O) through a respiratory chain of 3 large protein complexes, linked by 2 mobile carriers (co-enzyme Q and cytochrome C) on the inner mitochondrial membrane. In this sequence, NAD^+ (or $NADP^+$) and FAD are Vitamin B—derived co-enzymes. Q is a co-enzyme that consists of a substance *Ubiquinone,* which is a constituent of the mitochondrial lipids. CoQ has a structure that is very similar to Vitamin K and Vitamin E. It serves as a nighty mobile carrier of electrons between the flavoproteins and cytochromes of the electron transport chain likewise, cytochrome C is a small soluble protein that functions as a mobile electron carrier.

The electron carriers between CoQ and molecular O_2 are *cyto-chromes.* A cytochrome is an electron-transporting protein that contains a haem prosthetic group. The iron atom in cytochromes oscillates between a reduced ferrous (Fe^{2+}) state and an oxidised ferric (Fe^{3+}) state during electron transport. The *haem group is a one-electron carrier,* in contrast with NADH, Flavo-proteins, and CoQ, which are *two-electron carriers.* Therefore, a molecule of reduced CoQ must transfer its two high potential electrons to two molecules of cytochrome b, the next member of the electron transport, chain.

There are 5 slightly different cytochromes which operate in succession in the respiratory chain. They are named in order, cytochromes *b*, c_1 *c*, a and a_3. The terminal cytochrome a_3 *cytochrome oxidase*, is responsible for the final combination of reducing equivalents with molecular oxygen. It releases electrons back to H^+, which has remained in the reaction medium ; whole H atoms then combine with oxygen, forming water.

The salient features of the respiratory chain are as follows:

(1) After dehydrogenation of food, either NAD or NADP may accept hydrogen; depending upon the nature of specific foods. In the mitochondria of a cell, where most respiratory oxidations occur, it is usually NAD which serves as hydrogen carrier after food substrate.

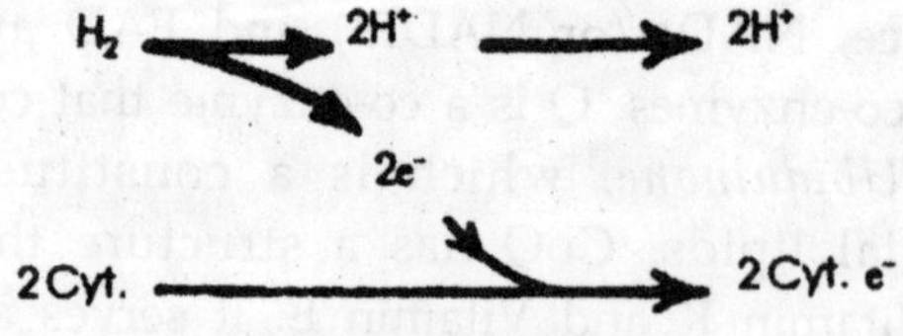

Transport of electrons by cytochromes.

NAD transports partly electrons, partly H atoms. Of each H_2 yielded by food, one whole H atom, and an electron from the other H atom join with electropositive NAD^+ (lacking one electron), resulting in neutral NAD. H. Thus NAD. H represents the reduced state and an H^+ ion remains in the medium. In hydrogen release, these reactions are reversed.

$$NAD^+ + H_2 \xrightarrow{\text{H ACCEPTING}} H^+ + NAD.H$$

$$NAD.H + H^+ \xrightarrow{\text{H DONATING}} H_2 + NAD^+$$

(2) FAD, co-enzyme Q and molecular oxygen transport whole H atoms only.

(3) All Cytochromes transport electrons, not whole H atoms. In this case, both H atoms of a pair dissociate

and each resulting electron is carried by one cytochrome molecule:

The H-carrying component of a cytochrome molecule is the iron in the centre of its tetrapyrrole ring. This iron normally exists in the form of ferric ion (Fe^{3+}). When it accepts an electron, its electropositivity is reduced by one charge unit and therefore, it becomes a ferrous ion (Fe^{2+}).

$$Fe^{3+} + e^{-} \longrightarrow Fe^{2+}$$

(4) The energy generated by the transfer of electrons through the respiratory chain is utilised in the pumping of protons (H^+) from the matrix side to the cytosolic side of the inner mitochondrial membrane. The H^+ concentration becomes higher on the cytosolic side and an electric potential with that side positive (+) is generated.

Peter Mitchell in his chemosmotic hypothesis (1961) proposed that a Proton Motive Force (PMF) is thus generated, consisting of (i) a *pH gradient* (cytosolic side acidic) and (ii) atrans-membrane electric potential (cytosolic side positive). Mitchell postulated that this PMF is used to drive ATP synthesis when protons flow back to the mitochondrial matrix through ATP synthase similarly as a flow of water from a higher to a lower level can be utilised to turn a water wheel.

Significance of Oxidative Phosphorylation

The NADH and $FADH_2$, formed in glycolysis, the citric acid cycle and fatty acid oxidation are energy rich molecules because they contain a pair of electrons which have a high transfer potential.

The concomitant transfer of electrons to molecular oxygen yields a large amount of energy which can be used to generate ATP. Once available, this energy can be utilised by the cell to drive all known endergonic process such as chemical syntheses, ion transport and mechanical exercise.

The generation of ATP molecules, as electrons are transferred from NADH or $FADH_2$ to O_2, by a series of electron earners is called oxidative phosphorylation. This is the major source of ATP in aerobic organisms. Some salient features of this process are:

(1) Oxidative phosphorylation is carried out by *respiratory assemblies* that are located in the *innermembrane of mitochondria.* A respiratory assembly contains numerous electron carriers such as the cytochromes. The reactions of the citric acid cycle and fatty acid oxidation, which supply most of the NADH and $FADH_2$, take place in the adjacent mitochondrial matrix.

(2) ATP is generated at three sites during oxidative phosphorylation as electrons flow through the Electron Transport System (ETS) from NADH to O_2. *Site 1* NADH dehydrogenase (complex I), *site* 2 is Cytochrome reductase (complex III); and *site 3* is Cytochrome oxidase (complex IV).

(3) Oxidation and phosphorylation are *coupled processes.* The oxidation of NADH yields 3 ATP, whereas the oxidation of $FADH_2$ yields 2 ATP because the electrons from this carrier enter the chain at CoQ, after the first phosphorylation site.

The rate of respiration in mitochondria can be controlled by the concentration of ADP. This is because oxidation and phosphorylation are *tightly coupled* and ADP is an essential component of the phosphorylation process. Thus electrons do not normally flow through the electron transport chain to O_2 unless ADP is simultaneously phosphorylated to ATP. In addition to ADP, oxidative phosphorylation requires a supply of NADH, inorganic phosphate and O_2.

Aerobic and Anaerobic Hydrogen Transfer

(1) *Aerobic:* Organisms with ETS system that use atmospheric oxygen as a final hydrogen acceptor are

considered aerobic (with air). It is the standard, universal form among animals.

(2) *Anaerobic* : Anaerobic means "in the absence of molecular oxygen". Whenever hydrogen transport from NAD to oxygenis blocked (due to cyanide poisoning, Vitamin B—deficiencies or inadequate supplies of atmospheric oxygen), organisms begin to respire in a way that does not require oxygen. This type of respiration is called *anaerobic* (without air) or *fermentation.*

With the path from NAD to oxygen blocked, pyruvic acid, one of the compounds normally formed in the course of carbohydrate degradation, readily accepts hydrogen. The fate of pyruvic acid thus serving as a hydrogen carrier may be determined along two lines:

(a) *Formation of Lactic Acid:* When pyruvic acid reacts with hydrogen, lactic acid is formed in animals. The reaction is catalysed by *Lactate dehydrogenase.*

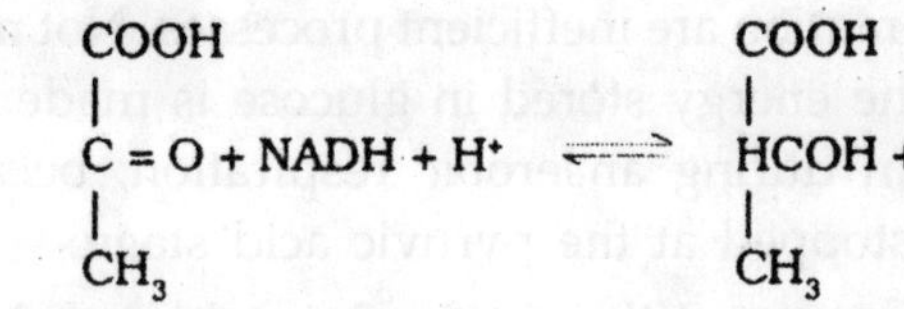

Pyruvic acid Lactic acid

Lactic acid then accumulates in the muscles causing fatigue and the animal must, therefore, stop intensive activity. During the later rest period, lactic acid is carried by blood from the muscles to the liver where it is converted to glucose. From the liver, glucose eventually returns via blood to muscles where it is transformed to glycogen. After the block to liver is no longer present or with extra oxygen now available, glycogen may be respired completely to CO_2.

(b) *Alcoholic Fermentation:* In plants and most micro-organism, pyruvic acid reacts with H_2 to form ethyl alcohol and CO_2. This process is called alcoholic fermentation. The great French biochemist Louis

Pasteur called fermentation "life without air". Pasteur found that fermentation was an energy-producing mechanism used by organisms deprived of air, like single celled fungi, the yeasts. The reaction takes place in 2 steps:

(c) In contrast to the reversible formation of lactic acid in animals, pyruvic acid in yeast is irreversibly decarboxylated to acetaldehyde by *pyruvic decarboxylase*, which is not present in animal tissues.

$$\underset{\textbf{Pyruvic acid}}{CH_3COCOOH} \xrightarrow{Mg^{2+}} \underset{\textbf{Acetaldehyde}}{CH_3CHO} + CO_2$$

(ii) The acetaldehyde thus formed is reduced by NADH in a reaction catalysed by *alcohol dehydrogenase.*

$$\underset{\text{Aceteldehyde}}{CH_3CHO} + NADH + H^+ \rightleftharpoons \underset{\text{Ethyl alcohol}}{CH_3CH_2OH} + NAD^+$$

Efficiency of Fermentation: Both lactic acid formation and alcoholic fermentation are inefficient processes. Not more than 5 per cent of the energy stored in glucose is made available to the organism during anaerobic respiration, because fuel oxidation has stopped at the pyruvic acid stage.

Potential energy still contained in this acid remains unutilised and locked in the end products of lactic acid or ethyl alcohol. Nevertheless, under conditions of oxygen debt in animals, such as during intensive muscular activity, fermentation can proceed alongwith aerobic respiration to provide a little extra energy.

Energy Transportation

Cellular oxidations take place with very small rise in temperature, because the energy produced as a result, does not accumulate primarily as heat. Instead, a large traction of this energy is tapped in the form of high energy bonds within ATP.

There are two types of chemical bonds in organic molecules: low-energy and high-energy bonds.

(i) *Low-Energy Bonds:* Most bonds in fuel compounds are of the *low-energy* type like C—C, C—H or C—O. When one of these is broken, only a comparatively small amount of energy is released.

(ii) *High-Energy Bonds:* Chemical bonds between certain atomic groupings in cells, such as phosphate groups, require large amounts of energy to be formed. Since they contain *extra energy,* correspondingly large amount of energy is released when they are broken. Although a *low energy* phosphate bond —O—P may also be formed that joins O and P but a wavy symbol ~ in the O ~ P links signifies the presence of a *high-energy* phosphate bond.

Conversion of AMP to ADP and ADP to ATP in biological systems requires additional phosphate groups and large energy inputs. Respiration provides this energy through fuel oxidation in cells, by which high energy bonds are created in ADP and ATP. ATP passes on this energy to drive those processes that require energy. Thus, ATP may be designated as the energy rich end-product of respiration.

The creation of high-energy phosphate bonds —O - P takes place in 4 steps:

(a) Phosphorylation of a fuel molecule by the attachment of a phosphate group -O-P.

(b) Dehydrogenation of phosphorylated fuel substrate.

(c) Dehydrogenation yields energy but since an -O-P group is now part of the fuel molecule, the energy is not released as heat. On the other hand, it becomes redistributed within the fuel molecule, and low-energy phosphate bond -O-P is transformed into high-energy bond —O~P.

(d) This high-energy bond is then transferred to ADP, generating ATP.

The ultimate source of phosphate groups in oxidation phosphorylations is the phosphoric acid made available to the cells as a mineral nu rient, but the immediate phosphate source in cells usually is ATP itself. For example, in the phosphorylation of glucose in cells, ATP is the donor.

Significance of the ADP—ATP High-Energy Cycle: ATP is the source of high energy phosphate for nearly every energy requiring reaction in the cell. The hydrolysis of 1 mole of ATP to ADP yields 7000 cal of (potential) energy which is used in various activities of a cell. In degradation reactions, the terminal high-energy phosphate of ATP is split off with the formation of a low energy phosphate and the conversion of ATP into ADP.

3

Nervous System

All multicellular animals contain elongated nerve cells, called *neurons,* collected together into parallel bundles called *nerve trunks.* Every nerve trunk contains thousands of neurons. The human central nervous system contains about 100 billion neurons. It is clear that these control the movement of muscles (via motor nerves) and the conduction of information from the sense organs to the central nervous system (via sensory nerves).

Framework of a Neuron

Each neuron has a cell body, called the *neurocyton,* 4 to 25 microns in diameter, containing a nucleus and cytoplasm, and from this radiate out one or more highly elongated processes. One type, called an *axon,* usually extends for some distance, and it is axons that make up the bulk of the visible nerve trunks. An *axon is the process of a nerve cell that carries impulses away from it.* The axon membrane is about 50 to 100 Å thick. The axon has many of the properties of an electric cable and is specialised to conduct the brief electrical waves called nerve impulses. The axons run parallel to one another, and each is surrounded along its whole length by a series of *sheath cells* (Schwann cells), which may have *myelin (medullated)* sheath, containing alternate layers of protein and fat molecules. In vertebrates, the larger axons have thicker myelin sheaths, and the axon proper may be up to 50 microns in diameter.

In addition to axon the cell body of a neuron gives off many smaller processes, which are called *dendrites* because they are usually highly branched like trees. A *dendrite or dendron is the process that carries impulses towards the cell body.*

Cell bodies and dendrites are usually aggregated into *ganglia,* and an animal's main assemblage of ganglia is called its *central nervous system* (CNS). Within ganglia the ends of axons come into very close proximity with cell bodies and dendrites of other neurons. These places of contact are called *Synapses.* At a *Synapse* the two cell membranes are usually somewhat thickened, and the gap between the two may be as small as 200 Å and hence only visible by electron microscope. A specialised type of synapse occurs where the axon of a motor nerve ends on a muscle cell.

Classification of Neurons

There are three common types of neurons: *motor, sensory* and *connector*. A sensory neuron has only a single process or fibre which divides close to the cell body into two main branches. Connector neuron consists of a cell body, one axon and numerous dendrites. Connector neurons occur exclusively in the CNS and serve as the immediate links between numerous sensory and motor neurons. The afferent or sensory neurons are located within the receptor organs or tissues, while the efferent or motor neurons are found in the spinal cord and their axons end in the effector organ, generally a muscle or gland which responds to the incoming nerve impulses. Thus, sensory neurons transmit the nerve impulses towards the CNS, whereas the motor neurons transmit them away from the CNS.

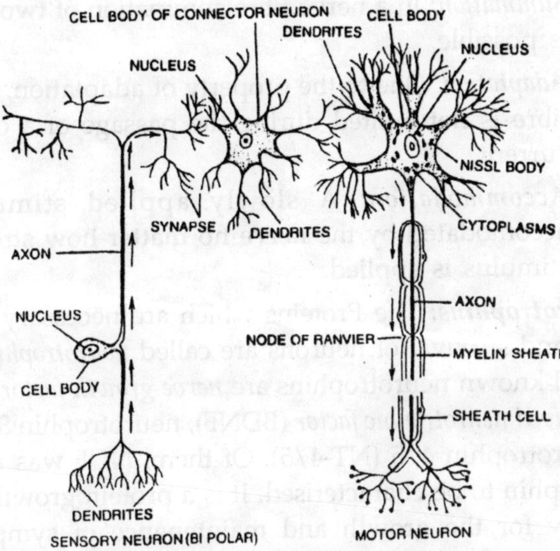

Three common types of neurons and the direction of nerve impulse transmission.

Functional Properties of the Nerve Fibres

1. *Excitability:* The nerve fibre can be stimulated by a suitable stimulus, which may be *mechanical, thermal, electrical* or *chemical* in nature.
2. *Conductivity:* The nerve impulse is conducted along the nerve fibre (normally in one direction only), and the conduction velocity depends upon the diameter of nerve fibre (the thicker fibres showing higher velocity) and temperature (in poikilotherms the velocity is lower than in nomeothermic animals).
3. *All-or-none-response:* A single nerve fibre will always give a maximum response to a suitable stimulus. If the strength or duration of the stimulus be further increased, no change in the response will occur.
4. *Refractory Period:* When a nerve fibre is once stimulated, it will not respond to a second stimulus for a brief period called the absolute refractory period.

5. *Summation:* In a nerve fibre summation of two stimuli is possible.
6. *Adaptation:* Due to the property of adaptation, a nerve fibre is not excited during the passage of a constant current.
7. *Accommodation:* A slowly applied stimulus is accomodated by the nerve no matter how strong the stimulus is applied.

Neurotrophins: The Proteins which are necessary for the growth and survival of neurons are called *neurotrophins*. The four well known neurotrophins are *nerve growth factor* (NGF), *brain-derived neurotrophic factor* (BDNF), neurotrophin 3 (NT-3) and neurotrophin 4/5 (NT-4/5). Of them, NGF was the first neurotrophin to be characterised. It is a protein growth factor necessary for the growth and maintenance of sympathetic neurons and some sensory neurons.

Neurotrophins bind to receptors (made up of two proteins that when dimerised have tyrosinkinase activity at the endings of a neuron. They are internalised and then transported to the neuronal cell body where they foster the production of proteins associated with neuronal development, growth and survival.

Physiology of the Transmission of Nerve Impulse: *The nerve impulse is that electrochemical change in the nerve fibre membrane that is aroused by a stimulus and is carried along the nerve fibre to its termination.* The transmission of the nerve impulse is a metabolic process, involving the consumption of O_2, CO_2 production, heat production, and the movement of ions. In very thin axons these impulses travel at less than one metre per second, but in large axons that activate muscles, they travel as fast as 100 metres per second.

Transmission in the Nerve Fibre: The transmission of nerve impulse has been explained by the so-called *membrane theory of nervous conduction*. According to this theory, the impulse itself is a transient set of biochemical reactions occurring mainly at the cell membrane of the nerve fibre (axon), and accompanied by a small and very brief change in membrane potential (Em), known as the *action potential*. However, the action potential is not the nerve impulse itself, it is only a means of showing that a nerve impulse is passing. The cell membrane of an inactive axon has a slight accumulation of Na^+ at its outer surface. Hence, the outside of the axon is electrically positive relative to the cytoplasm. This means that at rest there is a potential difference across the membrane of about 75 mV, (inside negative).

When such a *resting potential* exists, the membrane is said to be *polarised*. But when an impulse passes a given part of the axon, the cell membrane suddenly becomes much more permeable to positively charged ions than when it was at rest. Na^+ *ions then diffuse inwards,* and K^+ *outward* during the very small fraction of a second before the membrane recovers its resting properties and by means of active transport restores the previous distribution of Na^+ and K^+ ions. The influx of Na^+ ions causes *depolarisation* of the membrane and as impulse passes away, depolarised region becomes polarised again.

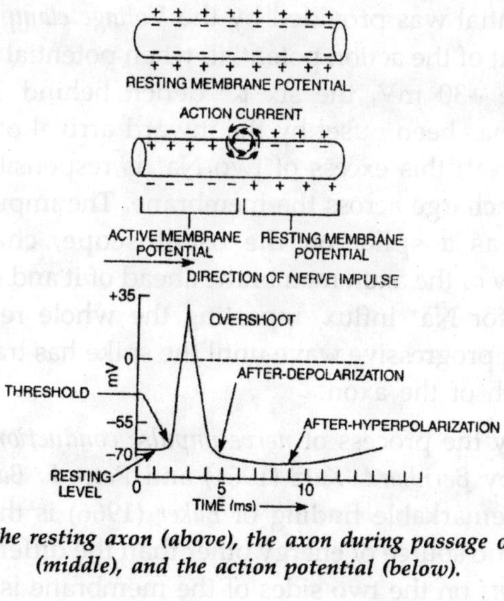

Showing the resting axon (above), the axon during passage of impulse (middle), and the action potential (below).

The electrochemical events that underlie the nerve impulse or action potential are determined largely by the axon membrane's differential permeability to Na^+ and K^+ ions and by the voltage difference across the membrane. As Na^+, ions leak through the membrane, they cancel out locally a portion of the excess negative charge inside the axon, thereby further reducing the voltage drop across the membrane. When the voltage drop has been reached to the *threshold* level, Na^+ ions enter in such numbers that they change the internal potential of the membrane from negative to positive. In other words, when potential difference of a small patch of membrane is lowered to its threshold, about -50 mV, the permeability of the membrane to Na^+ rapidly increases by as much as a hundredfold; Na^+ rushes across the membrane, in accord with the sodium gradient and attracted by the negative charge. This usually results in an *overshoot,* and the potential difference may go from -75 to +30 mV.

This process "ignites" and flares up to create the nerve impulse, or the *action potential* as it is called. The most convincing evidence for the movement of ions being responsible for the action potential was provided by the *Voltage clamp technique.* At the height of the action potential, when potential difference has become +30 mV, the six K^+ deficit behind 10 nm^2 of membrane has been offset by the inward arrival of a total of only eight Na^+; this excess of two Na^+ is responsible for the net positive charge across the membrane. The impulse which shows up as a spike on the oscilloscope, changes the permeability of the axon membrane ahead of it and sets up the conditions for Na^+ influx, repeating the whole regenerative process in a progressive wave until the spike has travelled the entire length of the axon.

Recently the process of *nerve impulse conduction* has been explained by *Bernhard 'Katz* (1961) and *Peter F. Baker* (1966). The most remarkable finding of *Baker* (1966) is the fact that apparently no source of energy other than the difference in *ion concentrations* on the two sides of the membrane is needed to amplify an electrical impulse and propagate it along the axon. During the passage of an impulse in an intact nerve there is small net gain of Na^+ and a small net loss of K^+. If this continues unchecked for sometime, the nerve would lose K^+ and gain Na^+, and the *concentration gradients* on which nerve activity depends would be destroyed. The tendency is counteracted by the mechanism known as the *"Sodium pump"*, which uses *metabolic energy* in the form of ATP to extrude Na^+ ions from the axon in exchange for K^+ ions. Although an intact nerve can function for sometime without an operating sodium pump, the pump is essential for the long-term maintenance of nerve activity.

Role of Saltatory Conduction

The impulse conduction in myelinated nerve fibres has been described by *Hoyle* (1965) as *saltatory conduction.* In non-medullated nerves the spread of depolarisation is continuous along the nerve fibre membrane and conduction is a uniform process. But in medullated nerves, in which myclin provides electrical insulation, the active generation of the current is confined to the nodes of Ranvier, and the impulse behaves as though it jumped from one node to the next (saltatory conduction). Actually, it does not "skip from node to node" but spreads along the internodal regions at a finite rate by cable conduction from amplifier to amplifier. Thus, only the membrane in the nodal region becomes active and the current flows from the next nearest resting node back over the outside of the myelin and returns in the core of the fibre.

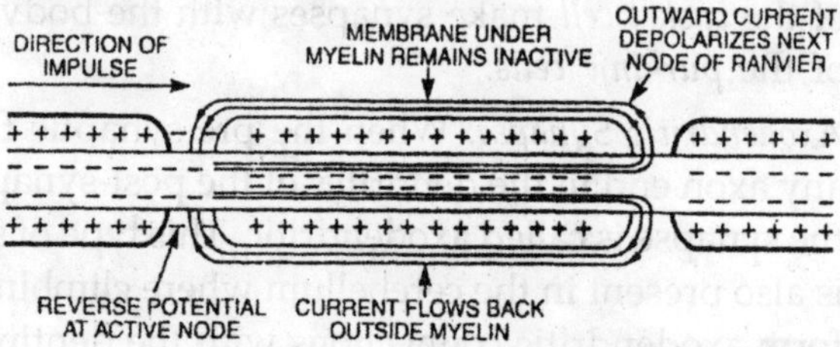

Saltatory conduction in myelinated nerve fibre.

Saltatory conduction is an evolutionary achievement of the vertebrates and possible of some higher invertebrates. This type of transmission confers two distinct advantages. Of them the first is the high conduction velocity a frog's medullated fibre of 20 μ diameter conducts at the same rate of 20m per second as a squid's giant axon of 500 μ diameter. The myelinated axons conduct up to 50 times faster than the fastest unmyelinated nerve fibre. The second advantage is one of metabolic economy. If only the nodes are depolarised, there is relatively less movement of ions across the membranes and naturally fewer demands on the "ion pumps.

Junction of Nerve Cells

The name "Synapse" was given by the Physiologist *Sir Charles Sherrington* (1906) to the junction point where the axon's terminal fibres make contact with another nerve cell.

The word synapse is derived from a Greek word meaning "to clasp tightly" *A synapse may be defined as the functional region where one neuron and other begins ; or synapse is the region where two neurons are functionally connected.*

Classification of Synapses: A synapse or *synaptic junction* is found in nerve-nerve and nerve-muscle interactions. According to the *nature of connection,* the synapses can be classified as:

(i) *Axosomatic Synapse:* When the pre-synaptic terminal of the axon ends in the cell body of the neuron, the synapse is called axosomatic, *e.g.* in the cerebellum the axons of the *basket cell* make synapses with the body (soma) of the *purkinje cells.*

(ii) *Axodendritic Synapse:* When the pre-synaptic fibres of any axon end in the dendrites of the post-synaptic cell, the synapse is called axodendritic. This type of synapse is also present in the cerebellum where climbing fibres form axodendritic connections with the dentine of the purkinje cell.

(iii) *Axo-axonal Synapase:* When pre-synaptic fibres of any axon end in the axon of the post-synaptic cells, the synapse of called axo-axonal.

According to the *nature of transmission* there are two types of synapses.

(i) *Chemical Synapses:* A chemical synapse is a junction between two nerve cells. The space between them is about 20 nm. In the nervous system of vertebrates transmission at the synapse is dependent upon the secretion and release of a *chemical* (neurotransmitter) which is released from *synaptic* vesicles, after they have fused with the nerve cell membrane into the *synaptic cleft* and is bound to receptors in the post-synaptic membrane This allows ions to flow through the receptor and causes the depolarisation of the post-synaptic cell membrane.

(ii) *Electrotonic Synapse:* In electrotonic synapse, a type of *gap junction*, the space between the cells is approximately 2nm. In this case no neurotransmitter is involved. The depolarisation of the pre-synaptic cell membrane during an action potential causes ion currents to flow through the gap junction.

In another type of electrical synapse, there is no gap between the cell membranes, which are fused to form a *tight junction.* The nerve impulse can be transmitted electrically across this tight junction often in either direction. Electrical synapses are far more frequently found in invertebrates than in vertebrates. Some vertebrates have electrotonic synapses in specialised areas ; for example, neurons controlling the electric organs of fish, the giant Mauthner cells at the base of the brain in tallest fish where transmission is electrical.

Functional Properties of Synapse

1. An impulse is allowed to pass through a synapse only *in one direction, i.e.,* from the axon of one neuron to the dendrite of the next. This is known as the *law of forward conduction.*

2. A synapse retards the passage of an impulse more than does a nerve fibre. Consequently, there is a *synoptic*

delay in the passage of an impulse from pre-synaptic to the post-synaptic fibre.

3. It is a site where impulses are received and discharged. However, a synapse not only acts as a *relay station* but it may also act as an integrator.
4. The physiological *seat of fatigue* probably lies in the synapses. The mechanism underlying the synaptic fatigue is pre-sumably due to exhaustion of transmitter material from *synaptic vesicles* following repeated pre-synaptic stimulation at a faster rate.
5. It has the characteristics of *summating* and *inhibiting* the impulses. Summation may be spatial.
6. Synapses bring about the *convergence* and *divergence* of nerve impulses.
7. It is very vulnerable to the action of many chemical compounds, some of which increase its responsiveness *(e.g., straychnine* blocks the inhibitory activity and causes reduction of inhibitory post-synaptic potential), whereas others decrease it *(e.g., anesthetics)*.

Structure of Neuromuscular Synapse: A neuromuscular synapse is formed when axon fibres (motor nerve or motoneuron) come in contact with the muscle cell (skeletal muscle fibres). The axon terminates in little knobs on the surface of the muscle fibre but does not fuse with it. Within the terminals of the axon are the spherical *synaptic vesicles* (40-200 nm in diameter) containing acteylcholine, and the many mitochondria needed for the active synthetic processes occuring in the terminals. Separating the axon terminal from the muscle fibre membrane is a gap of about 40 nm. This is called *synaptic cleft* and is wide enough to permit the flow of currents generated by the action of the chemical transmitter on the *post-synaptic membrane.* This membrane has junctional folds or post-synaptic folds (postsynaptic membrane thickining). The *junctional folds* contain particles which are believed to be acetylcholine receptors. The post-synaptic region responds only to chemical stimulation or inhibition.

Summation and Inhibition: Whether a given impulse crosses the synapse at which it arrives, is determined by two important factors called *summations* and *inhibition*.

Two or more impulses arriving at a given synapse are more likely than one to initiate another impulse in cell across the synaptic gap. They may arrive over the same axon within a small fraction of a second, in which case the phenomenon is called *temporal summation* or, they reach the synapses on a given cell over two or more different axons, in which case the process is called *spatial summation. A* common type of temporal summation is observed when a sensory axon entering the spinal cord sends impulses at different frequencies to a synapse on a motor neuron activating the muscle of a leg. Spatial summation often occurs at the same type of synapse when nerve impulses reach it over several sensory axons within a fraction of a second. Temporal and spatial summation are necessary for transmission of all impulses across synapses in the central nervous system.

Sometimes nerve impulses converging on a given synapse may produce just the opposite effect to summation. In that case one set of impulses may offset the effects of another, and this phenomenon is called *inhibition*. The higher the impulse frequency in an inhibitory nerve, the less will be the transmission across the synapse. Most cell bodies of neurons or their dendrites receive many synaptic endings where axons of other neurons can either excite or inhibit. Hence each neuron is always subject to both excitatory and *inhibitory* influence. It is the balance between these two influences that determines whether a given cell will itself be stimulated to convey the nerve impulse along its own axon. However, now it is known that a given axonal ending is either excitatory or inhibitory. But this is a property of the ending and not of the *post-synaptic neuron*. It is possible that many axons branch and lead to inhibitory endings at some synapse and excitatory endings at others. According to *Sir John Eccles* (1965), a nerve cell responds to the algebraic sum of the internal voltage changes produced by excitatory and inhibitory synapses.

Biochemical Mechanism of Synaptic Transmission and Role of Neuro transmitters: The transfer of nerve impulse across a synaptic gap is called synaptic transmission. All axons end at synapses. In the synapse a nerve impulse is transmitted from one nerve cell to another, or from *pre-synaptic* axonal endings to either the dendrites or the cell body of the *post-synaptic* cell. It has been shown that, when an impulse reaches an axon terminal, the terminal acts like a miniature endocrine gland. It secretes minute amounts of a hormone (or neurotransmitter substance) that diffuses through the synaptic gap or cleft. Some of it eventually reaches the dendrite terminals of the adjacent neurons, and the neurotransmitter there affects a dendrite in such a way that a new impulse is initiated in it.

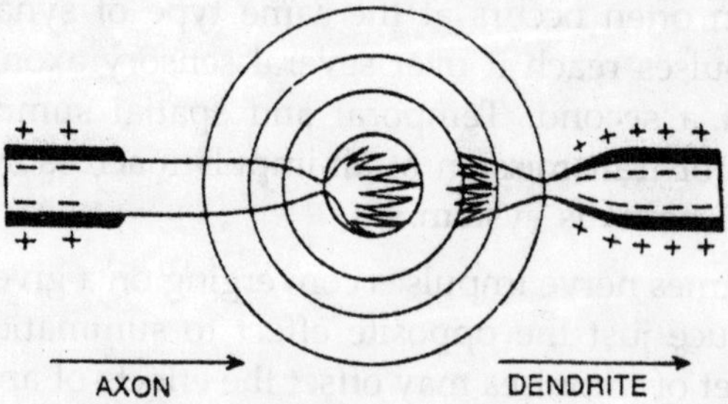

Diagram of a neural synapse, showing the release and local spreading of hormones (colour) from the axon terminal of one fibre to the dendrite terminal of another. Impulses are transmitted across the synapse by such chemical means.

In many instances the neurotransmitter is a *monoamine,* a substance the nerve cell synthesises by making minor changes in an amino acid. Norepinephrine, epinephrine, dopamine, and serotonin are *catecholamines,* compounds that contain a *catechol nucleus* (a benzene ring with two adjacent hydroxyl groups) and an *amine* group. These monoamine neurotransmitters are all derived from the amino acid *tyrosine.* Serotonin is an indolalkylamine derived from the amino acid *tryptohphane.* They are all classified as *monoamines* because they all possess a single amine group. Sometime the neurotransmitter is the amino acid itself (example, glutamic acid). In any case, its release and reception takes milliseconds (*Bloon,* 1981).

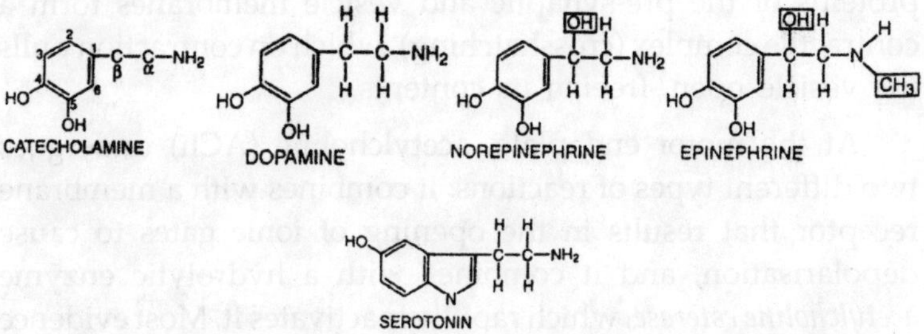

Structures of monoamines. Dopamine, norepinephrine, and epinephrine are catecholamines (compounds containing a catechol nucleus-a benezene ring with two adjacent benezene ring with two adjacent hydroxyl groups and an amine group). Serotonin is an indolamine.

In vertebrates, four neurotransmitter substances have been identified. They are *serotonin, acetylcholine, adrenaline* (epinephrine) and *nor adrenaline* (norepinephrine), and are collectively called *neurohumours,* meaning that they are produced by the nerve cells and stimulate or inhibit other cells. Adrenaline or noradrenaline is secreted by the axon terminals of the sympathetic post-ganglionic fibres, which are, therefore, said to be *adrenergic.* Serotonin or acetylcholine is produced by sympathetic pre-ganglionic fibres, all fibres of the parasympathetic system, and probably also by the fibres of the central nervous system. All such fibres are said to be *cholinergic.*

Gamma-aminobutyric acid (GABA) is the major inhibitory mediator in the brain where it helps in the transmission of about 20 per cent of synapses. It is also present in the retina and is the mediator responsible for pre-synaptic inhibition. Gaba transaminase (GABA-T) is the enzyme that catalyses the transmission.

Mechanism of Release and Action of Acetylcholine: The release of acetylcholine from synaptic vesicles is absolutely dependent upon the entry of Ca^{2+} which occurs when the action potential reaching the nerve terminals depolarises the membrane sufficiently (about 30 mV). One theory holds that the transmitter is released by *exocytosis* as the vesicle fuses with the pre-synaptic membrane. Another theory suggests that proteins of the pre-synaptic and vesicle membranes form a contractile complex (crosshatching), which on contraction pulls the vesicle open, freeing its contents.

At the motor end plate, acetylcholine (ACh) undergoes two different types of reactions: it combines with a membrane receptor that results in the opening of ionic gates to cause depolarisation, and it combines with a hydrolytic enzyme *acetylcholine esterase,* which rapidly inactivates it. Most evidence indicates that the ACh-receptor (AChR) is a protein consisting of 5 subunits surrounding a central channels (2a, and one each β , γ , δ) and a change in its structure occurs when ACh binds to it, resulting in the opening of ionic gates and a change in permeability.

Katz and *Miledi* (1967) estimated that there is a very high density of receptors (3×10^7) per end plate, which is enough for the 10^4 quanta of ACh liberated. There are from 12,000 to 21,000 molecules of ACh per quantam, which can easily be packed into a vesicle. Because the synaptic gap is very small, there is an excellent chance for most of the ACh molecules to colloide with a receptor site within a few milliseconds of their release and cause depolarisation within the requisite time for proper transmission.

In synapses formed by cholinergic fibres, the enzyme *acetylcholine esterase* (AChE) splits acetylcholine (ACh) into separate acetyl and choline fractions and so makes the hormone ineffective as shown below:

$$\underset{\text{Acetyl choline}}{CH_3CO{-}O{-}CH_2CH_2N(CH_3)_3} \quad \underset{\substack{-H_2O \\ \textit{choline acetylase}}}{\overset{\substack{+H_2O \\ \textit{acetyl choline esterase}}}{\rightleftharpoons}} \quad \underset{\text{acetic acid}}{CH_3COOH} + \underset{\text{choline}}{HOCH_2CH_2N(CH_3)_3}$$

The presence of this enzyme is advantageous adaptively. If it did not exist, ACh would linger in the synapse and would stimulate the dendrites repeatedly. If, therefore, two impulses arrived in a synapse in rapid succession, ACh produced by the first would still be effective when the second produced more hormone. The impulses would then merge into each other and the result would be one long drawn-out response, not two sharply distinct responses. Acetyl choline esterase prevents the merging of impulses. There is just time for ACh to stimulate the dendrites once or a few times and the hormone is destroyed immediately thereafter. An equivalent enzyme cannot be demonstrated in synapses formed by adrenergic fibres, the adrenalines do have a lingering effect. This circumstance contributes to the comparative slowness and sustained quality of the responses produced by the sympathetic part of the autonomic nervous system.

Drug Modified Transmission at Synaptic Junctions: Several drugs are able to modify transmission at synaptic junctions. *Muscarine,* an alkaloid, mimics the stimulatory action of ACh on glands and smooth muscle, although it has little effect on the receptors in autonomic ganglia. These muscarinic actions of acetylcholine involving muscarinic receptors are blocked by the drug *atropine.* In sympathetic ganglia, large amounts of ACh block transmission of impulses from pre to post-ganglionic neurons and this action, although unaffected by atropine, is mimicked by *nicotine.* As a result, these actions of ACh involving nicotinic receptors are called nicotinic actions. Both muscarinic and nicotinic ACh receptors occur in abundance in the brain.

Action of Acetylcholine

The *bacterial toxin of botulism,* the most poisonous substance known, prevents nerve endings from liberating acetylcholine and so killed by paralysis of the muscles controlling breathing.

Inhibitors of Acetykholinesterase: Acetylcholinesterase is located in the post-synaptic membrane. It is different from acetylcholine receptor and can be readily separated from it. Acetylcholinesterase does not bind *neurotoxins.* The toxic and therapeutic properties of acetylcholineasterase inhibitors are of considerable practical importance. Physostigmine, an alkaloid derived from the Calabar or ordeal bean, which was once used as an ordeal poison in witchcraft trials, inhibits acetylcholinesterase by forming a covalent intermediate that is hydrolysed very slowly.

However, acetylcholinesterase inhibited by organic phosphates such as di-isopropyl phosphofluoridate (DIPF) can be reactivated by derivatives of hydroxylamine (NH_2OH), which is used as a designed antidote for organophosphorus poisoning.

Inhibitors of Acetylcholine Receptor: Neuromuscular transmission can also be impaired by compounds that act directly on the *acetylcholine receptor* (AChR). For instance, *Curare* (one of the active components of *curare* is *d-tubocurarine),* a plant extract that South American Indians used to poison their arrow tips, blocks excitation of the muscle by binding to AChR on the sarcolemma and produces paralysis. It is used medically as a muscle relaxant during surgery.

Recently *α-bungarotoxin,* a protein molecule purified from snake venom, has been shown to bind on to a-subunits of the AChR and thus works as a specific receptor antagonist. Since the snake venom binds very tightly to the receptor, the estimation of how many receptor sites there are, can be made by counting the number of radiolabelled α-bungarotoxin molecules. A density of $3 \times 10^4 / \mu m^2$ has been observed in a neuromuscular junction in frog. The *Cobra toxin* also acts in a similar way.

Mechanism of Action of Catecholamines: *Julius Axelrod* (1974) has described the synthesis, release, mode of action, and the regulation of the neurotransmitters that are chemically known as the catecholamines (noradrenaline, adrenaline and dopamine). The enzymes involved in the synthesis of the transmitters are made in the cell body of the sympathetic neuron and transported down the axon to the nerve endings (studded with varicosities) where the transmitters are manufactured and stored in the vesicles.

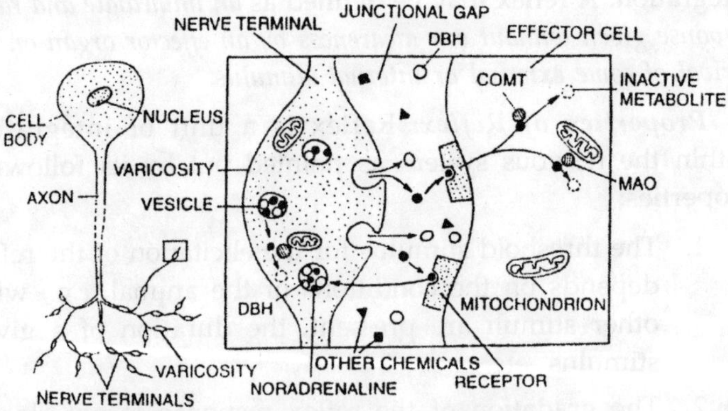

Diagram showing the release and inactivation of the neurotransmitter noradrenaline.

When the nerve is existed and depolarised, vesicles move to the cell membrane and fuse with it, releasing the neurotransmitter into the junctional gap. Once the neurotransmitter is liberated, it diffuses across the cleft between the nerve terminal and the adjacent cell. The capacity of the effector cell to respond to the transmitter then depends on the ability of its receptor to selectively recognise and combine with the neurotransmitter. When receptor and transmitter interact, a series of events is triggered that causes the effector cell to carry out its special function, including the propagation of the impulse across a synapse.

After interacting with the post- junctional cell, the adrenaline is inactivated by the enzymes catechol-O-methyltransferase (COMT) and monoamine oxidase (MAO) or more generally through its recapture by the pre-synaptic nerve terminals. The noradrenaline that leaks out of vesicles may be inactivated by the MAO stored in the membrane of the mitochondria.

Reflex and the Reflex Arc: The entire nervous activity is based on reflexes, the functional unit of the nervous system. *Sherrington* (1906) defined reflex as the unit of reaction in nervous integration because every reflex is a complete act of integration. A reflex may be defined as *an immediate and rapid response given without our awareness by an effector organ on the arrival of some external or internal stimulus.*

Properties of Reflex: Reflex is a unit of integration within the nervous system as pointed out by its following properties:

1. The threshold stimulus for the elicitation of the reflex depends on the conditions of the animal, *e.g.*, what other stimuli are pre-sent, the duration of a given stimulus, etc.
2. The gradation of the reflex response is not closely correlated with gradation of a stimulus above threshold.
3. Reflex action is usually the result of temporal summation. A single afferent impulse does not normally elicit a response.
4. There is little correspondence between the rhythm of a repetitive and the rhythm of the reflex response.
5. Most reflex actions are followed by a state of depressed excitability, which is often of long duration.
6. In many reflex actions there is an after-discharge phenomenon, *i.e.*, a prolongation of motoneuron activity after the stimulus has ceased.
7. The integrative action of a reflex is also shown by the fact that when two or more separate reflexes are stimulated simultaneously, one reflex is completely inhibited whereas the other proceeds normally. This is particularly the case when reflex actions are antagonistic.

Types of Reflexes: Reflexes are usually classified into two, the simple or unconditioned reflex, and the *conditioned reflex. A simple reflex is an inborn, inherited or unlearned response to a stimulus.* A familiar example of a simple reflex is the *knee-jerk reflex. A conditioned reflex is the response acquired as a result of training or experience to a stimulus that originally failed to evoke* *the reaction*. It was first clearly demonstrated by *Pavlov*. Under normal conditions the secretion of saliva when food is ingested is a simple reflex. *Pavlov* demonstrated on dogs that if an additional stimulus is regularly given (such as the ringing of a bell) at the time of giving the food to the animal, in time the mere ringing of the bell, without even seeing or smelling the food, would result in the secretion of saliva. Thus, a previously ineffective stimulus (the ringing of the bell in this case), by association with a stimulus (food) that evoked a response, had become effective in causing a response, the secretion of the saliva by the salivary glands. In other words, a conditioned reflex represents an inborn response which has been displaced to a new stimulus.

Reflexes are sometimes also classified as the *spinal reflex* and the *cranial reflexes*. The former are so-called because their basic neural path leads through the spinal nerves and spinal cord. Reflexes at the spinal level (involving skeletal muscle action) have the purpose of removing the animal from harmful stimuli and reflex actions are purposive and adaptive actions at the unconscious level. The cranial reflexes are executed by the brain and 12 pairs of cranial nerves. A large number of central nervous system reflexes are partly cranial, partly spinal. For example, usual stimuli may lead to muscular response in arms or legs. In such cases the sensory parts of the reflexes are cranial (via the optic nerves), and the motor portions are spinal (via the spinal nerves). Conversely, in other reflexes the sensory portions may be spinal and the motor portions may be cranial.

However, reflexes in the autonomic nervous system generally do not become conscious, irrespective of their course through the brain or the spinal cord. These reflexes regulate the functioning of body parts which are not under voluntary control, smooth muscles, heart, and most glands and other internal organs.

Reflexes may be phasic or tonic. The vertebrate knee jerk is a *phasic reflex* in which the muscle is quickly stretched and as quickly responds with a contraction to counteract the stretch. The nerve impulses that arise from the stimulus are quickly conducted to the CNS as a synchronous volley. The reflex efferent discharge from the CNS likewise passes a synchronous volley back to the muscle and causes a rapid contraction of all the fibres of the muscle.

The *tonic reflex* is characteristic, for example, of muscle activity used to maintain posture against the force of gravity. Thus, when a muscle is stretched slowly, varying numbers of stretch receptors are excited, only a few at first, but more as the muscle is stretched further or if the stretch is maintained. The result is an asynchronous discharge of impulses into the CNS and a resultant asynchronous discharge conducted along the motor nerve to the muscle fibres. As a result, a sustained contraction is produced in which, at any one time, only a relatively few muscle fibres are contracting.

The *myotatic reflex* depends on the activity of the stretch receptors in the muscles. Such reflexes are composed of both a phasic reflex and a static reflex. When the stretch on the muscle is brief, only the phasic reflex is seen (the best example of this is the knee jerk reflex). The peak of tension in a myotatic reflex occurs at the time of completion of the stretch movement (= phasic reflex action).

This is followed by a phase of static action represented by a plateau of gradually declining tension. The degree of tension maintained in the static phase is regulated by the degree of static stretch imposed on the muscle. The stretch reflex also has an inhibitory component, which can be demonstrated by recording the responses of an extensor muscle when its antagonistic flexor muscle is stretched. When stretched an extensor muscle shows a typical stretch reflex action. If a corresponding flexor muscle is stimulated while the extensor muscle is stretched, the stretch reflex does not occur. The stretch reflex plays an important role in walking.

The flexor or *withdrawal reflex* is the response evoked by the application of a painful or harmful stimulus to the limb of a spinal animal. This was called by *Sherrington* (1906) as *nociceptive reflex*. The *scratch reflex* results in a motor action in which the animal scratches its skin with a foot.

A simple example of scratch reflex in frog is seen when the animal is suspended so that the lower legs hang free. The application of small amount of dilute hydrochloric acid to the thigh evokes a reflex response in which the frog raises its foot in order to remove the painful stimulus. If the toes on the stimulated side are held so they cannot be moved, the opposite leg and foot are used to remove (scratch) the stimulating agent.

Reflex Arc: The structural and functional basis of the simple reflex is called the *reflex arc* and a *reflex action is the response to a stimulus carried over a reflex arc*. The most simple reflex arc consists of only two neurons, one sensory neuron and one motor neuron. But this type of reflex arcs are hardly met within higher animals. Most of their reflex arcs consists of more than two (at least three) neurons. A typical reflex arc consists of five parts or components: a receptor, an afferent or sensory neuron, modulator, efferent or motor neuron, and effector.

In a theoretically simple reflex arc consisting of a few neurons, the nerve impulse is normally initiated in the dendrites of the so-called sensory neuron located within the receptor organ. The impulses travel along the sensory neuron until they reach the axon endings and the synapse, through which the impulses are initiated in the second neuron in the arc, called the connector neuron. The connector neuron is always found in the CNS. The nerve impulse then proceeds along the length of the connector neuron, to its axon endings, and at another synapse initiates impulse in the third neuron, called the motor neuron. The terminal axon branches of the motor neuron end in effector organ, that is, in a muscle or gland which responds to the incoming nerve impulses.

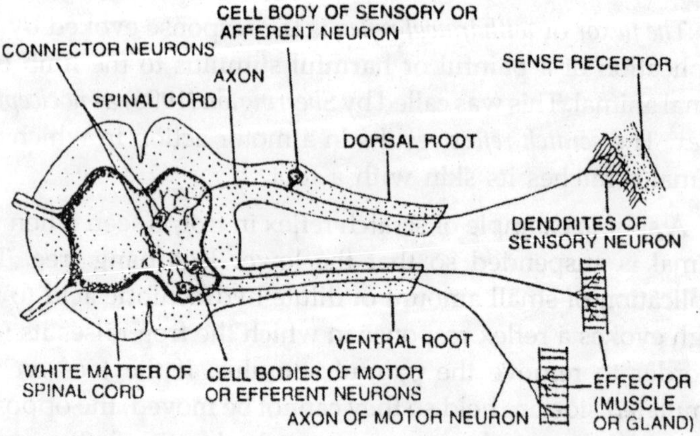

Showing the pathway of nerve impulse transmission in a reflex arc.

However, in higher animals a single stimulus may produce several simple reflexes, and their reflex actions almost always involve more than one reflex arcs. For example, should a person receive a sudden and unexpected burn on his finger tip, not only would he jerk his hand as a result of a simple reflex, but he would also respond in other ways, his behaviour and reactions involving a number of reflex arcs.

This happens because a single sensory neuron may have synaptic connections with other neurons, thus accounting for the ability of a single and highly localised stimulus to elicit several responses or reflexes. Similarly, a singled motor neuron may have synapses with numerous other neurons, thus accounting for the ability of different stimuli in different parts of the body to elicit the same response. As an example an injury to any part of the body may evoke the same use of the muscles of the larynx to emit a cry of pain.

Thus, most reflex actions are classified as complex and involve a number of neuronal chains. Complex reflexes involve several simple or even complex reflex pathways. Most reflex actions appear to be composed of chains of complex reflexive actions.

Reflex Centre: The term reflex centre refers to an assemblage of nerve cells (often including synapses) in the spinal cord or bixin which control the activity of a particular group of effectors, For example, the gray matter of the *medulla oblongata* contains several vital reflex centres including those that control the rate of heart beat, breathing, and constriction and dilation of blood vessels. Other reflex centres located in the medulla are those responsible for vomiting, coughing, sneezing, and swallowing, etc. The spinal cord also serves as an important reflex centre for great many reflexes. Other parts of the brain also serve as reflex centres, and in turn exercise influence on various body activities.

Nerve System

In the animal kingdom, the "higher" the organism, the more important becomes the system of cells that function to coordinate its activities. Nature has developed two distinct coordinating mechanisms.

One depends on the release and circulation of "chemical messengers", the hormones that are manufactured by endocrine glands that are capable of regulating the activity of cell in other parts of the body.

The second mechanism, which is in general far superior and selective, depends on a specialised system of nerve cells, or neurons, whose function is to receive and to give instructions by means of electrical impulses directed over specific pathways.

The nerve cells can be divided into two functional groups: sensory and motor. The sensory neurons collect and relay to higher centres in the nervous system the impulses that arise at the periphery of the body, *e.g.*, skin. The motor neurons carry impulses from the higher centres in brain and spinal cord to the "working" cells, stimulating muscular concentration or glandular secretion.

Functional Architecture of a Neuron: A *neuron* may be defined as a nerve cell consisting of a cell body which is drawn out into many processes. It is the morphological and functional unit of the nervous system. Neurons occur within the grey matter of the brain and spinal cord and in the ganglia of the cranial and spinal nerves and of the autonomic system.

A neuron consists of:

I. the cell body or perikaryon.

II. the processes of the nerve cell: an *axon* and *dendrons* or *dendrites.*

Cell Body or Parikaryon: The part of a complete neuron that looks like an ordinary cell is; the cell body (cyton or perikaryon). The cell membrane is a thin limiting border of condensed cytoplasm. The relatively unspecialised cytoplasm of the cyton is called neuroplasm in which *mitochondria* and *Golgi complex* are present in their typical form.

Neurofibrils are firm, protoplasmic threads, embedded in the neuroplasm. They exist in both the cell body and cell processes. Nissl bodies are small masses of granular material which have an affinity for basic dyes, owing to their content of ribonucleic acid. Nissl bodies are present in the cyton and branching dendrons but absent from most axons and their conical bases.

With the electron microscope, typical centriole can be seen in the cytoplasm, even though the neuron will never divide. Its role in the neuron is, therefore, very uncertain. Granules of pigment of two types can be distinguished in the neuroplasm. *Melaninis* present as dark brown or blank granules in certain cell groups. *Lipochrome* granules of a yellow or orange colour are found in the cells of autonomic ganglia, and more especially in the region of the axon hillock.

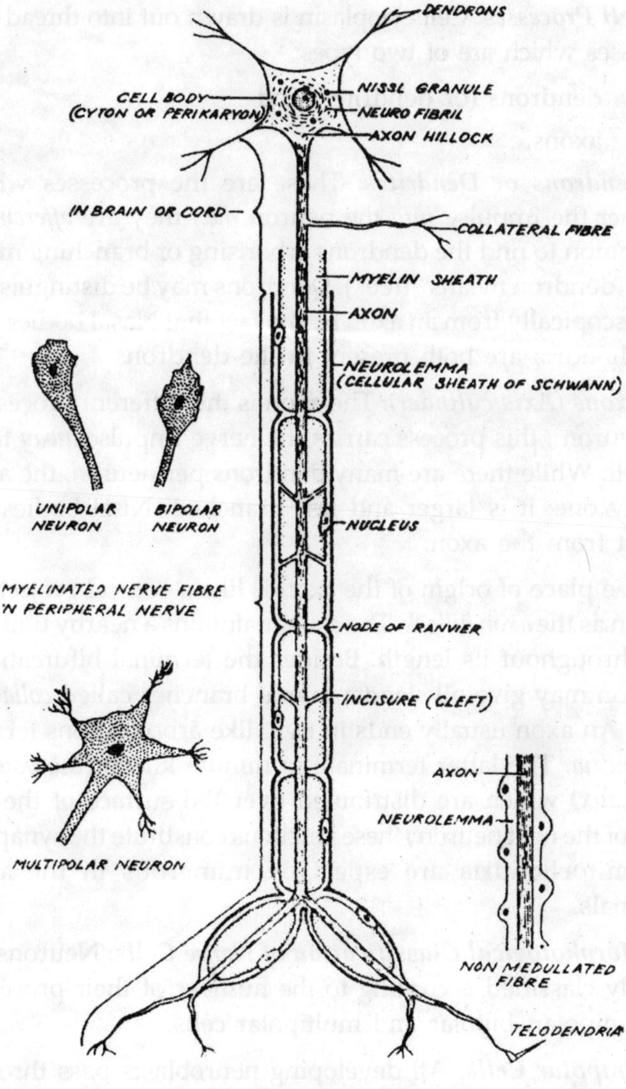

Nerve cells and fibres.

The nucleus is usually large, spherical and central, and varies in size with the size of the cell and its state of activity.

Cell Processes: Cell cytoplasm is drawn out into thread-like processes which are of two types:

(i) dendrons (or dendrites) and

(ii) axons.

Dendrons or Dendrites: These are the processes which conduct the impulse *into* the neuron, *i.e.*, they are *afferent*. It is common to find the dendrons arborising or branching many times (dendron means "tree"). Dendrons may be distinguished microscopically from an axon by the fact that Nasal bodies and mitochondria are both present in the dendrons.

Axons (Axis cylinder): The axon is the different process of the neuron ; this process carries the nerve impulse *away* from the cell. While there are many dendrons per neuron, the axon is only one; it is larger and less branched. Nissl bodies are absent from the axon.

The place of origin of the axon is like a conical projection, known as the *axon hillock.* The axon maintains a nearby uniform size throughout its length. Besides the terminal bifurcations, an axon may give oil slender lateral branches called *collateral fibres.* An axon usually ends in twig like arborisations termed *telodendria.* The latter terminate in minute knobs (the *boutons terminaux)* which are distributed over the surface of the cell body of the next neuron ; these junctions constitute the synapses. The mitochondria are especially numerous in the axon terminals.

Morphological Classification of Nerve Cells: Neurons are usually classified according to the number of their processes into unipolar, bipolar and multipolar cells.

Unipolar Cells: All developing neuroblasts pass through a stage when they have only one process, the axon. In the adult mammals such true unipolar cells are very uncommon, but they are found in the *mesencephalic nucleus of the V cranial nerve.*

Bipolar Cells: These cells are typically spindle-shaped, having the axon at one pole an a dendron at the other. Many developing neuroblasts pass through this stage, and in the adult they are found in the *reina,* in the *spiral ganglion of the cochlea,* in the *vestibular ganglion,* in the *olfactory neuro-epithelium.*

Multipolar Cells: These are nerve cells with many processes. Found in the *central nervous system* and *autonomic ganglia,* multipolar cells are the most common type of nerve cells. They may be pear-shaped, stellate or pyramidal in form.

Messenger Fibres

Many exons acquire coverings or *sheaths* along most of their course. An axon and its sheaths are commonly called a nerve fibre. Never fibres tend to run in bundles. Such aggregations are called tracts in the central nervous system and nerves in the peripheral system. Nerve fibres are of two kinds:

(1) *myelinated* and

(2) *unmyelinated.*

Myelinated (Medullated, White) Nerve Fibres: The white matter of the central nervous system, cranial and spinal nerves consist largely of myelinated fibres. These fibres consists typically of a central core *(axon)* surrounded by a tubular *myelin* sheath. The unspecialised cytoplasm of an axon is called *axoplasm.* The surface membrane of the axoplasm is known as *axolemma* which is directly concerned with the transmission of the nerve impulse. The myelin sheath is composed of lipoprotein; in the fresh condition it appears white and highly refractile. Electron microscope studies of Betty Ben Geren-Uzman and Francis O. Shmitt have revealed that myelin is a lamellar structure. The lamellae appear to be arranged concentrically around the axon like a jelly roll.

In peripheral nerve fibres the myelin sheath is interrupted at regular intervals; these interruptions constitute the *nodes of*

Ranvier, which appear as constrictions of the fibre when seen with the optical microscope. In the myelinated nerve fibre, resting and action potentials are limited to the nodes of Ranvier and this allows the type of nervous impulse transmission known as *saltatory conduction*. The myelin sheath is subdivided into *myelin segments* by oblique, circular *incisures* or clefts which arise as shearing defects. Myelin serves as an insulating jacket that improves the signalling efficiency of peripheral nerve fibres. The myelin sheath is lost near the peripheral termination of the fibre.

Myelinated nerve fibres of the cranio-spinal series are surrounded by another delicate sheath consisting of a series of separate, very much flattened nucleated *Schwann cells*, linked with each other. The part of a myelinated nerve fibre associated with one such cell, corresponds to an *internode, i.e.*, the distance between two successive nodes of Ranvier. In other words, the node of Ranvier is the junction between two Schwann cells. A single oval nucleus is located about midway between two nodes causing a small depression in the myel and a little outwards bulging of the neurolemmal membrane. The function of Schwann cells is to produce the myelin by winding its cytoplasm tightly around the surface of the axon. In this way, a spiral envelop (myelin) of many turns develops around the axon.

External to the Schwann cells and closely adhering to them is a very delicate connective tissue membrane, the neurolemma {or neurilemma), derived from the endoneurium of the nerve.

Non-myelinated (Grey) Nerve Fibres: These fibres are very much smaller and do not possess a myelin sheath. The axon is surrounded by a syncytial; nucleated sheath of Schwann cells. The axons of autonomic ganglia are of this type.

Origin and Propagation of Nerve Impulse

Nerve Impulse is the sum total of *chemical* and *physical events* in the propagation of a wave of physiological activity along a nerve fibre.

OR

It is a wave of *electrical depolarisation* and *repolarisation* sweeping along a nerve fibre.

Propagation in a Non-Myelinated Fibre: When a nerve fibre is at rest, *i.e.*, when it is not carrying an impulse, the outside of the fibre (surrounded by Na^+) is Positively charged in relation to the inside. The inside of the fibre is negatively charged even though it contain K^+ because of the presence of Cl^- and a variety of LARGE organic anions (resulting from metabolism), such as acetate, pyruvate, lactate and amino acids - that are too large to diffuse easily through the axon membrane.

Extracellular Fluid (ECF)			***Intracellular Fluid***
CATIONS	Na^+	145	12
	K^+	155	
ANTON'S	Cl^-	120	3.8
HCO_s-	27	8	
	A-	7	155
POTENTIAL		0	—90 mV

1. *Origin of Resting Potential:* There is a differential distribution of ions between the axoplasm and the external medium. The extra cellular fluid (EFC) has a high concentration of Na^+ and Cl^- and the Intracellular fluid a high concentration of K^+ and of larger organic anions (A^-).

A potential difference arises when a membrane is selectively permeably to either anions or cations, and a difference in concentration of some non-diffusible ion exists between the inside and the outside of the membrane. Then a potential develops between the two sides, and the *solution of higher concentration becomes negative to the one of lower concentration.* This difference in electric potential on the two sides of the nerve fibre membrane is termed the membrane potential or resting potential which is about — .07 volt or — 70 mV. The nerve fibre membrane is said to be polarised in this state. The resting potential depends on a physico-chemical equilibrium condition known as the Donnan Equilibrium which predicts the movements of Na^+, K^+ and Cl^- ions into and out of living cells.

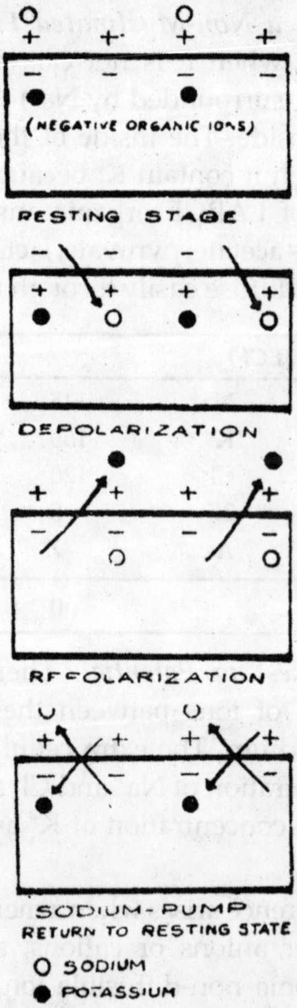

Movement of ions at a single point on the nerve fibre.

2. *Depolarisation:* Stimulation of a nerve changes the permeability of the nerve fibre membrane. The stimulus may be of mechanical, electrical or chemical nature. As

a result Na^+ quickly moves in. Since each Na^+ ion is positively charged, this movement results in a positive charge on the inside of the cell as compared with the outside. This reversal of electric charges is called depolarisation.

3. *Repolarisation:* K^+ ion with their positive charge, now begin to flow out faster than the Na^+ flows in. When these positive charges are moved to the outside of the cell, they initiate the process of replorisation (return to the resting state and inside of the cell again becomes negative.
4. *Metabolic Pump: The active transport* of Na^+ out of the cell and the re-establishment of K^+concentration inside it restores the cell to a resting state. This mechanism is called Sodium Potassium Exchange Pump because it transports sodium out of the cell and at the same time also moves Potassium back into the cell, *against concentration, gradient by expenditure of energy derived* from cellular metabolism.
5. *The Action Potential:* As Na^+ rushes into the fibre, the outside of the fibre briefly becomes negatively charged at that point. Adjacent to this point, a part of the fibre remains positively charged. This difference in potential between the two areas is called the Action Potential. This is a temporary change between a stimulated and resting portions of a nerve fibre.

Since opposite charges attract one another, positive charges move towards the negative charges on both sides of the membrane. The result is a *partial neutralisation* of the charge or a small *decrease* in the electric potential (a *depolarisation)* of the membrane adjoining the site of the impulse. This depolarisation causes the membrane to become more permeable. As a result Na^+ rushes into the cell ; the process ignites' and flares up to create the nerve impulse or action potential. Thus, action potential may be defined as a self propagating depolarisation of the nerve membrane. Bit by bit, the impulse moves down the fibre. This is a regenerative process that leads to automatic self-reinforcement. The nerve impulse is thus a self-propagating wave of depolarisation followed by repolarisation moving down the nerve fibre.

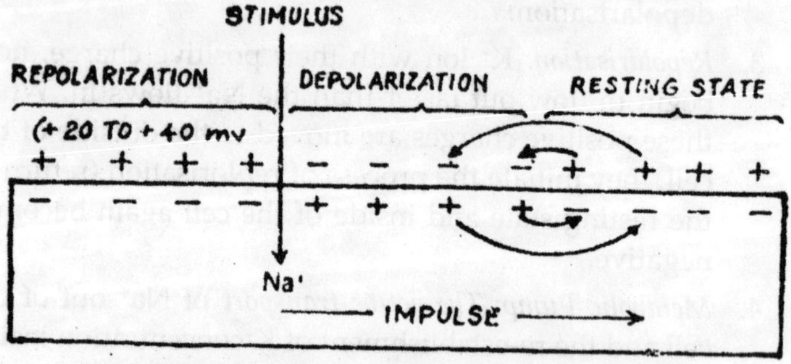

Local current low during the passage of an impulse along an un-myelinated nerve fibre.

The nerve impulse may be compared to the ignition of a string of fire-crackers. When a burning match stick is applied to the end of the string, its threshold is reached, and a spark is generated which begins to pass along the string. The energy required for this movement is derived not from the energy of the burning match stick, but from the chemicals contained in the fire-crackers. The spark does not weaken with the distance travelled. If the string branches, the spark passes along each branch with undiminished strength.

The nerve impulse travels much more slowly than an electric current and by a different means:

Electric Current and Nerve Impulse

Electric Current	*Nerve Impulse*
1. When an electric current travels along a conductor, the conductor serves only as a passive carrier of the electrons that move in the same direction as the current, because this	1. During transmission of a nerve impulse ions are the only particles that move. Moreover, they move at right angles to the movement of impulse, i.e., they move

Contd...

Electric Current	Nerve Impulse
movements of electrons is the current.	in the out of the fibre.
2. An electric impulse is conducted along the wire at the speed of the light.	2. In the neuron, an electrochemical reaction is simply moving down the length of the neuron.

Although a nerve impulse is not an electrical current flowing along a nerve fibre, the impulse is accompanied by measurable changes in electrical potential in the membrane.

Propagation in Myelinated Nerves: Myelin is a lipid-rich wrapping of high electric resistance arid functions as an effective insulator. But this insulation is broken as the nodes of Ranvier. Accordingly the current (Action Potential) flows from one node over the myelin to the next node. Because, the charges leap from node to node, this form of propagation is termed Saltatory Propagation in which the current spreads by *Jumping, skipping, dancing* or *leaping* from one node to another instead of travelling along the nerve membrane.

Saltatory propagation increases the speed of the impulse, because only the membrane at the nodes of Ranvier must undergo depolarisation. As a result, impulse transmission is about 20 times faster in a myelinated than in an unmyelinated neuron.

Velocity of Propagation: It ranges from a few metres/sec for thin non-myelinated fibres to about 100 metres/sec in the thickest myelinated fibres. For example, in large mammalian nerve fibres. Highest speeds, equivalent to some 200 miles/hour are found in the sensory and motor fibres concerned with body balance and fast reflex movements in non-myelinated axons, the greater the fibre diameter, the lower is the internal resistance and higher the conduction velocity. In myelinated fibres, the velocity is found to be directly related to diameter, because intemodal distance is proportion to the diameter.

Synapse and Synaptic Transmission

Structure: There is always more than one neurone involved in the transmission of a nerve impulse from its origin to its effector organ, whether it is a sensory or motor pathway.

The junction between two neurons is known as a synapse. Actually synapse is a gap, the synaple cleft, between the end of one nerve fibre and the beginning of the next. In most cases *post-synaptic neuron* forms terminal swellings or synaptic knobs on the *post- synaptic neuron.* The size of the synaptic knob varies from 0.5 to 2 μ in diameter. The synapic knob contains numerous synaptic vesicles which are preformed "packets" of a chemical transmitter substance - the *acetylcholine.* This is formed in the cell body from where it is transported down the axon to the pre-synaptic terminals. The formation of acetylcholine is helped by a special enzyme, *choline acetylase.*

Action of Acetylcholine: The arrival of an impulse at the synaptic knob causes acetylcholine to be discharged into the synaptic cleft. Accumulation of acetylcholine in the synaptic gap reduces the polarisation of dendrites, or cell body of the next, *follower,* neuron. If depolarisation is sufficient, the threshold of the next neurone will be reached and a nerve impulse generated in it. In this way, acetylcholine serves as a *chemical transmitter* of the nerve impulse.

Removal of Acetylcholine: As nerve impulse after nerve impulse causes the secretion of acetylcholine into the synapse, it becomes highly concentrated and diffuses in all directions. But as a matter of fact, it never gets very far or becomes very concentrated. An enzyme, *acetylcholinesterase* is present in the vicinity of syriapases which breaks down the acetylcholine thereby preventing continued depolarisation of the next fibre or the inordinate spreading of the nerve impulse to other nerve fibres.

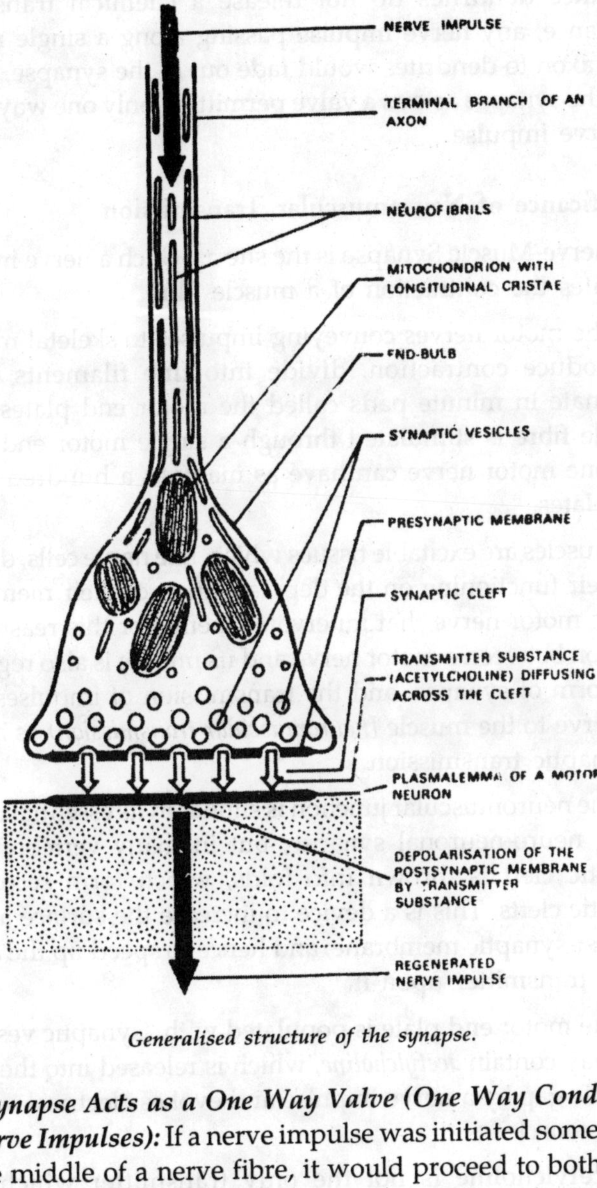

Generalised structure of the synapse.

Synapse Acts as a One Way Valve (One Way Conduction of Nerve Impulses): If a nerve impulse was initiated somewhere in the middle of a nerve fibre, it would proceed to both ends. However, in the synapse, the propagation is strictly one-way, from the axon to the dendrite-never the reverse.

Since dendrites do not release a chemical transmitter substance, any nerve impulse passing along a single neuron from axon to dendrites would fade out at the synapse. In this way the synapse acts as a valve permitting only one way traffic of nerve impulse.

Significance of Neuromuscular Transmission

Nerve-Muscle Synapse is the site at which a nerve impulse activates the contraction of a muscle fibre.

The motor nerves conveying impulses to skeletal muscles to produce contraction, divide into fine filaments which terminate in minute pads called the motor end plates. Each muscle fibre is stimulated through a single motor end plate, and one motor nerve can have as many as a hundred motor end plates.

Muscles are excitable tissues which, like nerve cells, depend for their functioning on the depolarisation of their membrane by the motor nerve that innervates them. For this reason, the junction between a motor nerve and its muscle is also regarded as a form of synapse, and the transmission of impulses from the nerve to the muscle *[neuromuscular transmission)* as a form of synaptic transmission.

The neuromuscular junction is similar in its essential details to the neuro-neuronal synapse, but in many vertebrates its synaptic cleft is thrown into folds, which form secondary synaptic clefts. This is a device to increase the surface area of the post-synaptic membrane, and hence to speed up the action of the transmitter upon it.

The motor end plate is populated with "synaptic vesicles" that may contain *acetylcholine,* which is released into the short synaptic gap by a nerve impulse and evokes electrical activity in the muscle.

Acetylcholine is not the only transmitter which may promote transmission across excitatory synapses. For example, *noradrenaline* acts as a transmitter at the neuromuscular junctions of mammalian smooth muscle, and at certain central synapses. *5—hydroxy tryptamine (serotonin)* and *gamma aminobutyric acid (GABA)* are other neurotransmitters of physiological importance.

Functional Properties of a Nerve Fibre Threshold: The stimulus applied to a nerve fibre must reach a certain minimum strength or threshold before it can initiate an action potential to travel along the nerve fibre.

The All or None Response: Once the threshold level has been attained or exceeded, the impulse is found to reach its maximum height independently of any further increase in the intensity of the applied stimulus.

Once started, the nerve impulse travels at a speed that is entirety independent of the intensity of the stimulus. A nerve impulse is, therefore, referred to as an all-or-non phenomenon: it requires a minimum intensity of stimulus of elicit it, but once initiated, it will proceed to its maximum size.

Refractory Period: The frequency with which impulses can pass along a nerve fibre is governed by the refractory period of the fibre. If an action potential is initiated by an applied electric shock, a second stimulus applied during the period when the action potential is reaching its peak, will not produce a second action potential, however intense the stimulus. This period of complete inexcitability is called the *absolute refractory period.*

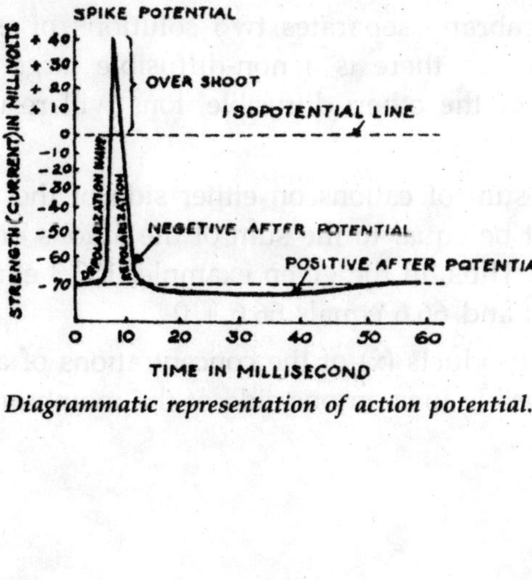

Diagrammatic representation of action potential.

The absolute refractory period determines the maximum frequency of conduction. In mammalian nerves the observed refractory period ranges from 1 to 6 milliseconds, which means that approximately 200 to 1000 impulses per second can be fired along a nerve.

Spike Potential: If the neuron is activated by an adequate stimulus, a marked alternation in the potential takes place. There is a very rapid change in potential from the resting level of — 70 mV to' zero (Isopotential Line) and then to + 20 or 40 mV. With almost equal suddenness, the potential then returns to the resting level.

The sharp rise and fall in the potential constitutes the spike potential. The reversal of polarity is called the overshoot.

Negative after Potential: There is a short section of the action potential just before it reaches the resting level where the rate of fall becomes abruptly slow. This slower fall is called the *negative after potential.*

Positive after Potential: During replorisation, the potential returns to the resting level and in most cases, *beyond, i.e.,* to about — 75 mV. The segment of the curve that falls below the resting level is known as the *positive after-potential.*

Donnan Equilibrium and its Characteristics

If a membrane separates two solutions of electrovites in one of which there is a non-diffusible large on (P^-), a distribution of the other, diffusible, ions will result. This is because:

(1) The sum of cations on either side of the membrane must be equal to the sum of the anions on the same side. Thus, in the given example, 133.3 equals 33.3 + 100 ; and 66.6 equals 66.6 + 0.

(2) The products (x) of the concentrations of any pair of diffusible ions on one side of the membrane must be

equal to the product of the concentrations of the same pair of ions on the other side. Thus, 133.3 x 33.3 equals 66.6 x 66.6.

The net result is that an equilibrium condition becomes established in which there remains some imbalance of ions on either side of the membrane, and the P^- side of the system is more negative than the other, *i.e.*, there is a potential difference across the membrane.

Electrical effect of the P^- ions will be not only to draw in positive ions, but to repel the Cl^- ions to an equal extent. At equilibrium, therefore, the ratios of Na^+ ions on the P^- side to those on the other side will be the reciprocal of the corresponding ratio for Cl^- ions.

Donnam Equilibrium

Initial State				*At Equilibrium*	
Side 1		Side 2		Side 1	Side 2
Na^+	P^-	Na^+	Cl^-	Na^+ Cl^- P^-	Na^+ Cl^- P^-
100	100	100	100	133.3 33.3 100	66.6 66.6 0

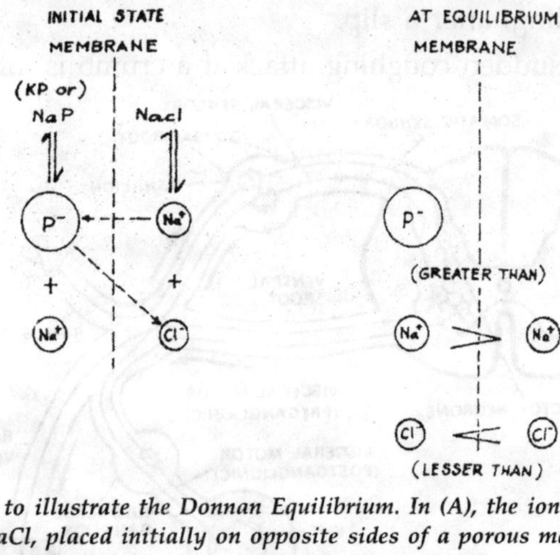

Diagram to illustrate the Donnan Equilibrium. In (A), the ionisation of NaP and NaCl, placed initially on opposite sides of a porous membrane, are shown, the electrical forces exerted by the P-ions membrane, are shown. The electrical forces exerted by the P- ions are indicated by broken lines. The final equilibrium situation is shown in (B).

The Reflex Action and the Reflex Arc: Most of our body activities and reactions to environmental changes, both internal and external, are automatic, or in voluntary. Such responses which occur outside of our awareness, are immediate and rapid and are called reflexes. A reflex action can be defined as an automatic motor response to a sensory stimulus without the brain being immediately involved. Reflexes are usually classified in two broad subdivisions: (i) the *simple, of unconditioned reflex,* and (ii) the *conditioned reflex.*

Simple Reflex: A simple reflex is an inborn, unlearned response to a stimulus or change in environment. Most reflex actions are protective in function. There are a number of familiar examples.

1. The quick jerk of the leg if the patella tendon is sharply tapped ; this is termed the *knee Jerk.*
2. The quick closing of an eyelid when an object suddenly approaches the eye. This is known as the *corneal reflex.*
3. The rapid withdrawal of a hand when it is burned or pricked.
4. The quick recovery of the balance of the body to prevent falling, after a slip.
5. A sudden coughing attack if a crumb is inhaled.

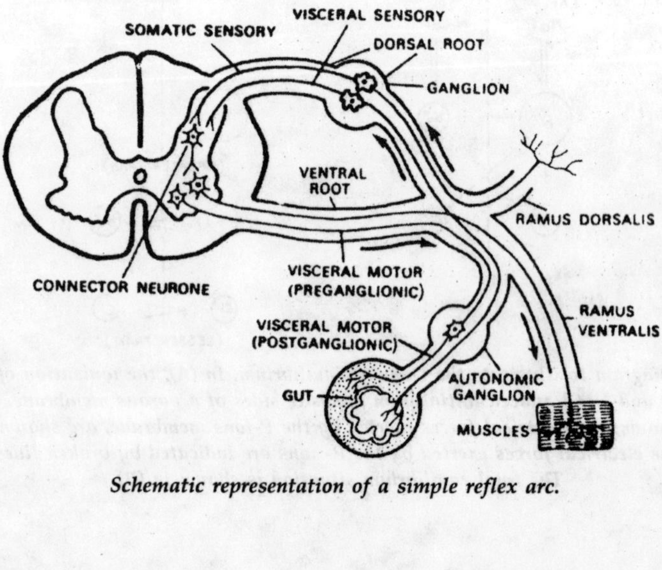

Schematic representation of a simple reflex arc.

Many other reflex actions occur within the body which never reach consciousness, for example, glandular secretions, movements of the stomach and small intestine, the changes in the calibre of blood vessels, breathing, rate of heart beat, and so on.

Reflex Arc: The anatomical basis of the simple reflex, called the reflex arc in an arrangement of atleast three neurons in a pathway that always passes through the central nervous system. The three components are as follows:

1. A sensory neuron which comprises the *sensory nerve endings* in a receptor organ, the *sensory nerve, posterior (dorsal) root ganglion cells and its nerve fibre* which passes to the dorsal horn of grey matter in the spinal cord.
2. A connector neuron which consist of a *nerve cell* its *dendrites* and *axon* in the spinal cord.
3. A motor neuron consisting of a *nerve cell* and its *dendrites* in the ventral horn of the grey matter. The *axon* of this nerve cell and the *motor end plates* terminate in the effector organ, *e.g.*, muscles.

Physiology of Reflex Action: In a simple reflex, the nerve impulse is initiated when the sensory nerve endings in the receptor organ are stimulated. The impulse travels along the *afferent* or sensory neuron until it reaches the terminal branches of the axon which form a synapse with the dendrites of the connector neuron. In the spinal cord, the impulse proceeds along the length of this connector neuron to its axon endings which form another synapse with the dendrites of the motor neuron.

At this synapse, an impulse is initiated in the *efferent* or motor neuron. The terminal axon branches of the motor neuron end in the effector organ. The reflex or response made by the effector organ as a result of the arriving nerve impulse is either muscular contraction or glandular secretion, depending on the effector involved.

Coordinating Properties of the Reflex Arc: Reflex actions of higher animals almost always involve many reflex arcs, not just one. For example, if an individual receives a sudden, unexpected burn out his fingertip, he would not only jerk his hand as a result of a simple reflex, but would so respond in other ways. He would turn his head to the source of danger, cry out and very likely experience a series of sensations and reactions including pain, fear, and emotional tension.

The total reaction, therefore, would involve numerous muscles — including those of the arm, shoulder, neck, head trunk, tongue, farynx and respiratory tract. Besides, several mental processes participate in such an integrative response. Thus, a simple stimulus, *i.e.*, a mild, highly localised burn, evolves an extensive reflex behaviour involving the coordinated response of the entire nervous system.

Inhibitions of Reflex Actions: In the infant the act of micturition and defecation is carried out by reflex action and there is no voluntary control over these processes. As the child grows older, the sensory nerve tracts of the spinal cord become fully developed ; the impulses from the bladder and rectum reach the brain and can be controlled. Thus the act of micturition and defecation becomes a conscious voluntary act in the adult.

Many reflex actions are inhibited by the body, thus preventing the overactivity of the muscle which otherwise would contract in response to every sensory stimulus, even although a response was unnecessary.

Certain reflex actions cannot be inhibited by the cerebrum, for example, the secretion of glands, vaso motor, respiratory, cardiac and alimentary reflexes.

If there is damage to the spinal cord in an adult due to injury or disease, the stimuli would not reach the sensory area of the brain. As a result, the individual will not be conscious of a full bladder or rectum. When the pressure reaches a certain level the bladder or rectum will be emptied by reflex action. This involuntary action is called incontinence of urine and faeces.

Reflex Centre: The term reflex centre refers to a group of neurons in the spinal cord or brain that controls the activity of a particular group of effectors. For example, the reflex centre for breathing located in the *medulla oblongata* of brain controls the impulses conducted to thoracic muscles responsible for breathing action. The cells of the reflex centre are in turn regulated by appropriate afferent impulses. Similarly, the reflex centre for heart-rate controls and adjusts the heart beat by motor impulses in response to sensory massages from aorta or the vena cava. Other reflex centres located in the medulla are those responsible for *vasodilation—vasoconstriction, vomiting, coughing, sneezing and swallowing.*

Conditioned Reflex: Conditioned reflexes, in contrast to simple reflexes, are not inborn but are acquired and important on past experience, training, and learning. The conditioned reflex was first clearly demonstrated in the early part of this century by the Russian physiologist Ivan Petrovitch Pavlov (1849— 1936). He showed that conditioned reflex can be developed experimentally to an extraordinary degree. Pavlov present a hungry dog with food, which elicited a reflex of salvation by the stimulation of the taste buds in the mouth. He then rang a bell each time the dog was given the food in order to associate the unconditioned response (salivation) with an additional stimulus (the bell). After a few trials, Pavlov was able to demonstrate that the dog salivated at the sound of the bell, without the sight or smell of the food. This is known as a conditioned reflex.

4

System of Reproduction

The reproduction may be defined as a self-directed process by which every kind of living organism produces new individuals of its own kind. There are two main types of reproduction: *asexual* and *sexual*. Sexual reproduction is of universal occurrence among all higher forms, including man. However, it has certain biological advantages and disadvantages over the asexual method. One of its advantages is that sexual reproduction permits the combining within a single individual of the best genes of its parents. The chief disadvantages of the sexual method are the hazards involved in the meeting of eggs and sperms and the possibilities of unfavourable growing conditions. But the fact that the sexual process is so extensively distributed in different organisms in such a vast variety of forms suggests that it serves in a role of broad biological significance in addition to its function in reproduction.

In both sexes, the gonads have a dual function: the production of *germ cells* (sperms by testes and ova by ovaries) and the secretion of sex *hormones* (androgens by testes and estrogens by ovaries.

Reproductive System of Males

The male reproductive system in man consists of primary sex organs, the testes, and a set of accessory sexual structures.

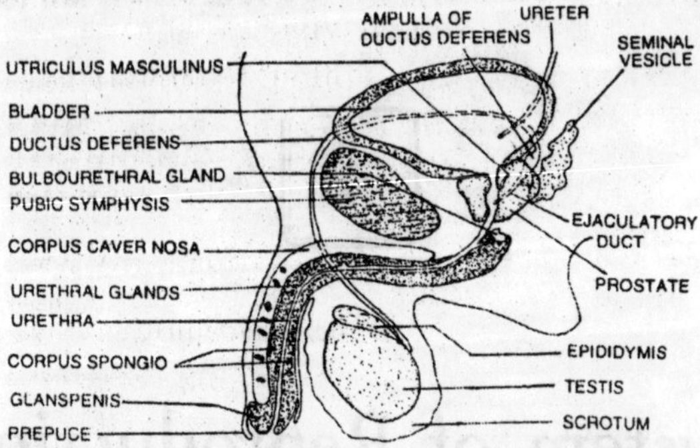

The male reproductive organs in median view (after Eberth).

Testis: Each of the two testes in the adult male is an ovoid body about the size of a walnut. Both are located outside the body cavity within a skin covered pouch called the *scrotum* or *scrotal sac.* The two principal functions of the testes are the formation of sperm cells by *spermatogenesis,* and the production of male sex hormone, *testosterone.*

The wall of each testis consists largely of connective tissue which is continuous with wall like partitions dividing the interior of the testis into approximately 250 small compartments called *lobules*. Each lobule contains from one to three tiny convoluted *seminiferous tubules* within which the spermatozoa are formed. *Sertoli cells* are believed to regulate the differentiation of spermatozoa and to nourish them. The special-cells situated among the seminiferous tubules are collectively called the *interstitial tissue,* which performs the important function of producing and secreting testosterone.

Function of Hormones

Both the structure and function of the testes are dependent upon hormonal stimulation by the pituitary *gonadotrophins,* which in turn are regulated by hypothalamic secretions. The hypothalamus of male secretes at least one, and perhaps two, gonadotropin-releasing hormones (GnRH): *luteinising hormone-releasing hormone* (LHRH) and *follicle-stimulating hormone-releasing hormone* (FSHRH). These hypothalamic peptide hormones cause the release of LH and FSH from the anterior pituitary which perform the following functions.

(i) *Regulation of Testostreone secretion by LH*: The most important function of LH in the male is to stimulate the interstitial cells of Leydig to secrete *testosterone.* In the absence of LH (also known as ICSH or interstitial cell-stimulating hormone) the interstitial cells undergo severe atrophy and cease functioning. The testosterone, an androgenic steroid hormone, is essential for puberal Fig. Testosterone, development and structural and functional maintenance of the male accessory reproductive organs.

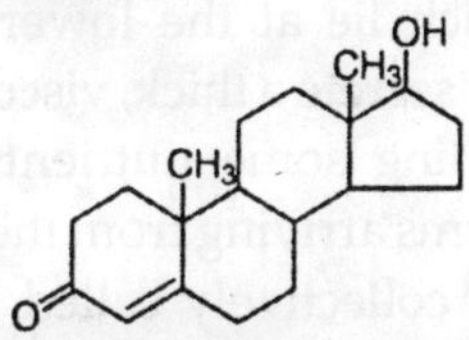

Testosterone.

(ii) *Regulation of Spermatogenesis by FSH:* The maintenance of the structure and the spermatogenic function of the seminiferous tubules is dependent upon the action of the *follicle-stimulating hormone* (FSH). It has been shown to bind specifically to the Sertoli cells, indicating the importance of FSH to spermatogenesis. However, FSH is known to stimulate spermatogenesis but cannot complete the maturation process into spermatozoa unless testosterone is present. Therefore, LH and FSH must be secreted simultaneously by the anterior pituitary for proper spermatogenesis to occur.

Accessory Structures of Male Reproductive System: The accessory ducts and glands of the male reproductive system include *vasa efferentia, epididymis, vas deferens, ejaculatory ducts,* *penis* and *glands*. The *prostate gland* secretes by way of numerous ducts leading into the urethra, an alkaline fluid which is believed to activate or increase the motility of the sperm and to neutralise the acid condition of the urethra and the female reproductive tract. In old man the prostate gland frequently becomes enlarged, thus constricting the urethra and making urination difficult.

Surgical removal of the gland is often used as a remedy for this condition. Slightly below the prostate gland is a pair of small glands (each about the size and shape of a small pea) collectively called the *Cowper's glands* or *Bulbourethral glands*. The mucous secretion of these glands drain into the urethra, serving as a lubricant. They also help to counteract acidity as does the prostatic secretion.

The paired *seminal vesicles*, are hollow glands about two inches in length which lie at the lower back surface of the urinary bladder. They secrete a thick, viscous, slightly yellowish seminal fluid containing some nutrients. The seminal fluid together with the sperms arriving from the testes and secretions by others glands are collectively called the *semen*. However, the secretory activity of seminal vesicles is dependent upon the action of testosterone.

The penis or the male copulatory organ consists of *erectile tissue* through which the urethra runs. During sexual excitement the sponge-like spaces of the erectile tissue become engorged with blood under pressure due to dilation of the tissue's arteries and constriction of its veins. The resulting turgidity causes the penis to become enlarged, hard and erect. In the sexual intercourse (also called mating, coitus, or copulation) stimulation of the sensory nerve endings in the penis as a result of friction against the lubricated walls of the vagina culminates in a short period of heightened paroxysmal sexual excitement called the *orgasm*.

The orgasm is accompanied almost simultaneously by *ejaculation*, the forceful coming out of semen from the penis.

However, ejaculation is the result of several coordinated involuntary muscular contractions including those of the walls of the epididymis and vas deferens, propelling the sperms from the epididymis through the ducts into the urethra. The average volume of semen ejaculated during coitus is about 2.5 to 3.5 millilitre. There are normally about 100 million sperms per ml of semen. About fifty per cent of men with counts of 20-40 million per ml and essentially all of those with counts under 20 million per ml are sterile. Sperms are usually released in the upper region of the vagina near the cervix, and are moved principally by muscular contractions of the female reproductive tract into the uterus and then into the oviducts.

Reproductive System of Females

The reproductive system of human female consists of two female gonads (the ovaries) which produce the female gametes or egg cells, and associated ducts and structures. The latter include the oviducts (also called egg ducts or Fallopian tubes), uterus, vagina, and vulva. The female reproductive tract is concerned with: (1) the reception and transport of eggs and spermatozoa, making their union possible, (2) physiologic accommodation of the fertilized ovum and resulting foetus, and (3) the forcible delivery of the foetus at term.

The female sexual organs normally remain in an infantile state during approximately the first ten years of life. In the succeeding 2 to 4 years there occurs a gradual enlargement of the reproductive organs, accompanied by the growth of the breasts, changes in the contours of the body, and the appearance of axillary and pubic hair. This period of sexual development culminates in the appearance of the first menses *(menarche)* at an average of 14 years. Mature reproductive life, characterised by continually recurring menstrual cycles, lasts until about 45th to 50th years of life. The human female then enters a period known as the *menopause* during which the reproductive cycles become irregular, lengthen, and eventually cease.

Thenceforth, in the *post-menopause*, the reproductive organs are functionless.

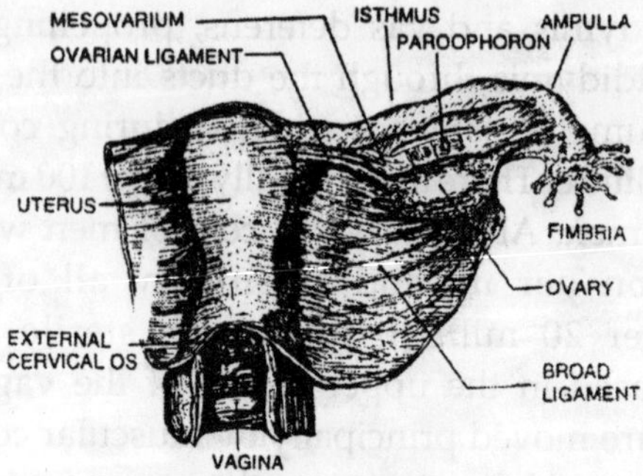

Reproductive organs of human female.

Role of Ovary

Each of the two adult ovaries is about the size and shape of a large almond and is separately located in the pelvic region of the abdominal cavity. The surface of the ovary consists of a layer of columnar epithelial cells which at various intervals penetrate deeply as columns of cells into the underlying connective tissue. During embryonic development of female small groups of cells separate from these columns and become arranged into definite structures known as the *primordial* follicles, each consisting essentially of a large single cell surrounded usually by a single layer of cells.

At the time of birth, there are 2 million ova, but 50 per cent of these are atretic. The atresia continues during development and the number of ova in both the ovaries at the time of puberty is less than 300,000. Only one of these ova per cycle (or about 500 in the course of a normal reproductive life) normally reaches maturity; the remaining degenerate.

Growth of Follicles and their Functions

The process that leads to discharge of an ovum from the ovary involves great increase in the diameter of the follicle. When this developmental process has been initiated the follicles are referred to as the *primary follicles.* The latter are 30 to 40 μ in diameter, whereas the diameter of the mature follicles may reach 10 to 12 mm. This increase in size is produced both by proliferation of the granulosa cells and by accumulation of the

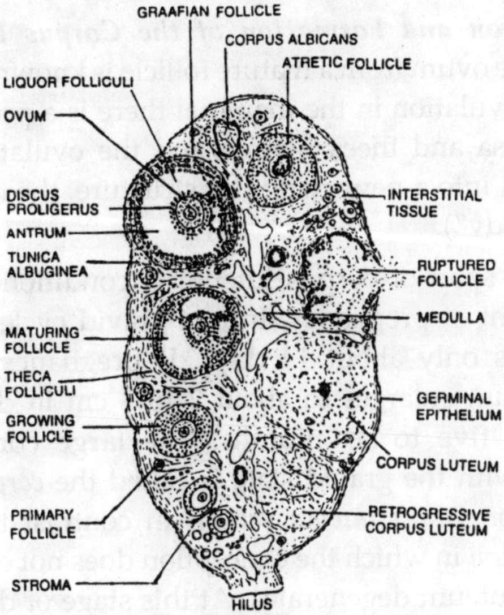

Mammalian ovary. Progressive stages in the differentiation of a follicle are indicated on the left side.

liquor folliculi within the follicle. At the initiation of follicle growth, the follicle cells begin to multiply, forming an increasing number of layers of cells. At the same time the ovum enlarges through the accumulation of yolk substances within its cytoplasm, and a new membrane, the *zona pellucida*, is deposited on the surface of the ovum.

As the follicle expands in diameter, it sinks deeper within the cortex and the neighbouring ovarian stroma are molded about it, forming the *theca folliculi.* The latter differentiates into two layers, the *theca internet* and the *theca externa.* When the follicle wall becomes six to nine cell layers thick, small lakes of follicular fluid form and almost immediately begin to coalesce, forming a crescent-shaped vesicle, the *antrum.* As the antrum enlarges through further accumulation of liquor, the ovum is displaced eccentrically, and the granulosa is disposed as a thin layer of cells, the *stratum granulosum,* lining the fluid-filled antrum. After the accumulation of liquor the follicle is termed as the *vesicular follicle* or the *Graafian follicle.*

Ovulation and Formation of the Corpus Luteum: The escape of the ovum from a mature follicle is known as *ovulation.* Following ovulation in the mammal there is a proliferation of the granulosa and theca interna and the ovulated follicle is transformed into a new glandular structure, the *corpus luteum* ("yellow body").

The life span of corpus luteum is conditioned upon the establishment of pregnancy. In non-gravid cycles the corpus luteum lasts only about 14 days. If pregnancy occurs, the corpus luteum enlarges to about 2 to 3 cm in diameter and persists for five to six months. The large corpus luteum associated with the gravid state is called the *corpus luteum of pregnancy* (or *corpus luteum verum)* in contrast to the *corpus luteum spurium* in which the conception does not occur. Finally the corpus luteum degenerates. At this stage of degeneration, the corpus luteum has a whitish colour and is termed as the *corpus albicans.*

Comparable morphological bodies are formed in the ovaries of many other vertebrates, ranging from the cyclostomes to the mammals. They may arise during follicular atresia (pre-ovulatory corpora lutea) or after ovulation (post-ovulatory corpora lutea). After many years of controversy (*Hoar* 1969), now it seems likely that these bodies produce hormones in some of the fishes and reptiles as well as in the birds and mammals.

Importance of Endocrine Functions of Ovary

As in males, the gonadotropic hormones secreted by the adenohypophysis, *i.e.,* FSH and LH, control the formation and growth of the follicles. The ovaries, in addition to producting germ cells, secrete the sex hormones *estradiol* and *progesterone.*

Estradiol is produced primarily by the maturing follicles, and progesterone primarily by the corpus luteum. Estradiol induces growth and development of the female reproductive tract and the mammary glands. Progesterone or the so-called pregnancy hormone stimulates the uterine glands to secrete, renders the *endometrium* (the internal layer of the wall of the uterus) receptive to the ovum, inhibits the spontaneous contractions of the uterine musculature, and promotes breast development.

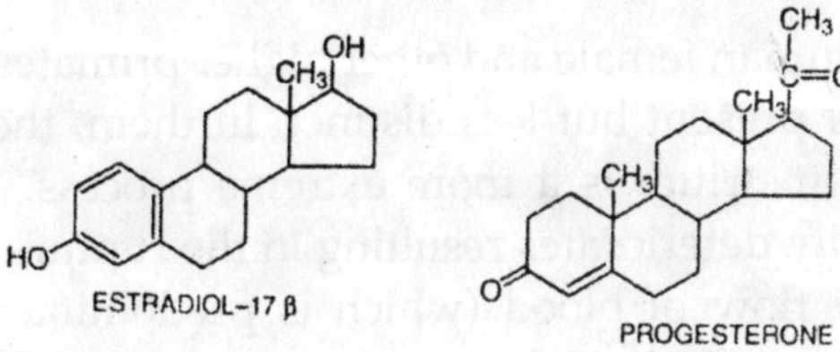

Structure of the important female sex harmonu

Importance of Estrous Cycle

The relatively brief periods within the breeding season when the female of the most mammalian species (with the exception of higher primates, including man) undergoes spontaneous ovulation and is most receptive of the male is known as the "heat" or *estrous*. The rhythmic or cyclic changes in the development, release and maturation of the egg cell, together with the accompanying structural and physiological changes in the reproductive system as well as behavioural changes, periodically prepare the female for mating, and are collectively termed the estrous cycle.

However, the endometrium undergoes a marked transformation during the estrous cycle. At the beginning of the estrous the relatively thin endometrium gradually thickens, its blood supply increases, and its uterine glands deepen as the follicular maturation proceeds. With the release of the egg cell and the subsequent appearance of the corpus luteum, the endometrium continues to thicken and build up even more rapidly. If fertilization occurs, the zygote implants itself in the endometrium. If fertilization does not occur the egg cell degenerates, the corpus luteum begins to disappear, and the endometrium regresses returning to its original thin state. The endometrium does not build up again until the beginning of the next estrous cycle. If pregnancy is absent the estrous cycle will be regularly repeated according to the mating pattern of the species.

Importance of Menstrual Cycle

In the human female and other higher primates the estrous cycle is still present but less distinct. In them, the regression of the endometrium is a more extreme process. The uterine lining rapidly deteriorates resulting in the rupture of its blood vessels, the flow of blood (which is predominantly arterial, only 25 per cent being of venous origin) and other fluid, and the sloughing away of most of the endometrium, phenomena collectively called the *menstruation*.

The cycle is thus uniquely marked by regularly occurring periods of menstruation. The period of normal menstruation (menses) which lasts about 4 or 5 days, occurs between the periods of ovulation and therefore, does not correspond to the estrous. The cycle is accordingly called *menstrual cycle* and in human female averages about 28 days from the start to the end, before the next cycle begins. Some women who are infertile have *anovulatory cycles, i.e.,* they do not ovulate but have menstrual periods at fairly regular intervals.

Changes in the endometrium of the uterus during the menstrual cycle are shown in Fig. The growth of the superficial and middle layers of endometerium occurs from the 5th to the 14th day of the cycle under the influence of *estrogen* produced by the growing ovarian follicle. At approximately this time ovulation occurs and the follicle is transformed into the progesterone-producing corpus luteum. In the second part of the cycle, following ovulation, when *progesterone* is the dominant sex hormone, a tremendous increase in growth and coiling of the superficial arteries occurs in preparation for the increased blood supply necessary for an implanting blastocyst. The stroma becomes loose and edematous, which facilitates implantation.

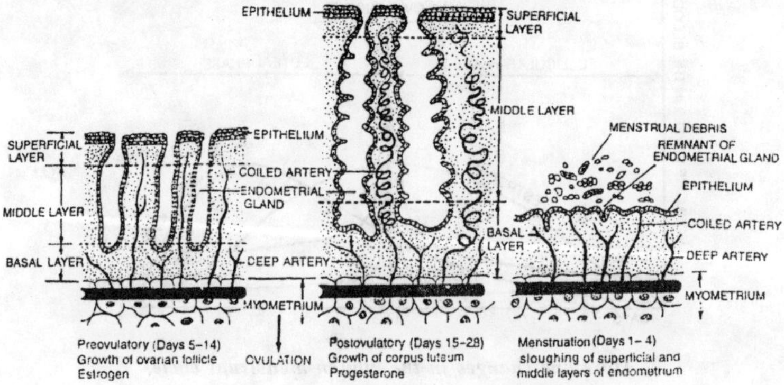

Changes in the endometrium of the uterus during the menstrual cycle.

The endometrial glands proliferate and become secretory, producing large quantities of glycogen and mucin. It has been suggested that these substances may be helpful in allowing the blastocyst to adhere to the endometrium and that the glycogen may be used as nourishment for a short while. Progesterone also slows the contractions of the myometrium, so that the blastocyst has a chance to adhere and implant itself in the richly vascular tissue that has been prepared so carefully for it. The implantation leads to a suppression of the menstruation.

Hormonal Regulation of Menstrual Cycle: The first two weeks of a menstrual cycle are dominated by FSH, estrogen and the follicle. This period is called the *follicular phase* and terminates with ovulation. The second two weeks are dominated by LH, progesterone and the corpus luteum. This period is the *luteal phase* (secretory phase). If pregnancy does not begin, during the *luteal phase,* menstruation takes place in the course of the first few days of the next menstrual cycle. A new cycle starts when, in the absence of LH and progesterone, the pituitary resumes FSH production and a new follicle matures in the ovary. Thus FSH and estrogen are the components of one control cycle, LH and progesterone of another ; and these two occur alternately, the termination of one being the stimulus for the initiation of another.

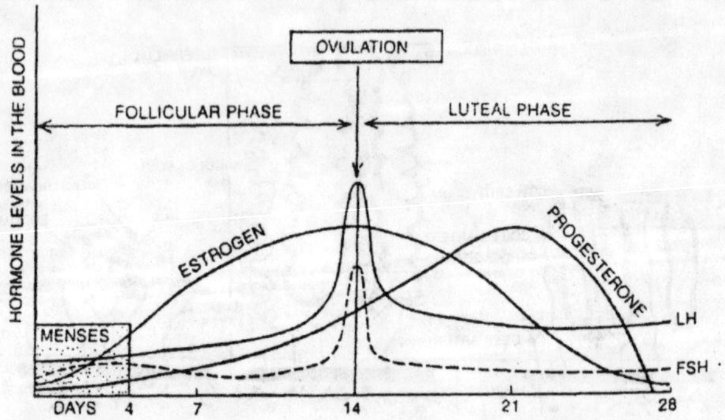

Hormone changes in the human menstrual cycle.

However, we may note that estrogen is not completely absent during the luteal phase. Some can be shown to be present in the blood even then, despite the absence of follicles at that time. Similarly, some progesterone is present during the follicular phase of the menstrual cycle. Exactly where these hormones come from is still undetermined. Immature follicles which will attain maturity only during future cycles might secrete some estrogen during a given luteal phase.

Analogously, corpora lutea from the past cycles and in the process of degeneration might secrete some progesterone during a given follicular phase. Or other ovarian tissues might secrete the hormones. Yet, although both female sex hormones are present at every stage of the menstrual cycle, their quantities do fluctuate sharply. Estrogen reaches a definite peak late during the follicular phase and at the time of ovulation, whereas progesterone reaches the peak concentrations during the luteal phase of the cycle.

This fluctuation in the concentrations of estrogen and progesterone has far-reaching consequences. In addition to its effect on the uterus, progesterone promotes the development of the duct system in the mammary glands. It induces differentiation of estrogen-prepared ductal tissue and supports the secretory function of the breast during lactation. Body temperature increases somewhat during the follicular phase, then falls during the luteal phase. Sex drive is likely to be more pronounced during the follicular phase, since estrogen maintains it. Since estrogens, like androgens in males, affect mental processes, it is possible that the monthly fluctuation in these hormones contributes to the emotional fluctuations rather characteristic of the females.

Impregnation and Implantation

In humans, the time of greatest fertility coincides roughly with the time of ovulation. The fertilization of the ovum by the sperm usually occurs in the midportion of the uterine tube. Sperm binds to a sperm receptor called ZP_3 in the zona pellucide, a membrane that surrounds the ovum and this is followed by the acrosomal reaction, the breakdown of the acrosome (the lysosome like organelle on the head of the sperm). During the process some enzymes are released, including the trysin-like protease *acrosin*. Acrosin facilitates the penetration of the sperm through the zona pellucida. The fusion of sperm with the membrane is mediated by *fertilin,* a protein on the surface of the sperm head that resembles the viral fusion proteins which permit viruses to attack cells. The fusion provides the signal that initiates development.

The developing embryo, now called a *blastocyst,* moves down the tube into the uterus. This journey takes 3 to 4 days during which the blastocyst reaches the 8 or 16 cell stage. After coming in contact with the endometruim, the blastocyst is surrounded by an outer layer of *syncytiotrophoblast,* a multinucleate mass with no cell boundaries, and an inner layer of *cytotrophoblast* made up of individual cells.

The syncytiotrophoblast erodes the endometrium and the blastocyst burrows into it. This is called implantation. The implantation site is usually on the dorsal wall of the uterus. A placenta then develops and the trophoblast remains associated with it.

Gestation and its Application

The pregnant female contributes vitally to the offspring development. The most extensive maternal contribution is made while the offspring is retained within the environment of the uterus, *i.e.,* during the period of *gestation.* A final contribution is made when the course of pregnancy terminates, during the process of birth.

(a) *Luteal Phase of Gestation:* When the zygote reaches the uterus a few days after fertilization, it is actually already an embryo composed of several hundred cells. This embryo is deposited on the endometrium which has been readied for its reception by the progesterone. The embryo next becomes *implanted* in the wall of the uterus, *i.e.,* it is gradually surrounded by the uterus tissue. Such implantation leads to a suppression of the menstruation. Through still poorly identified nervous pathways, the presence of an embryo in the uterus wall is signalled *via* the brain to the pituitary which continues to produce LH as a result. The corpus luteum consequently continues to secrete progesterone. Thus the luteal phase of the menstrual cycle in which fertilization has occurred now does not terminate as usual but continues directly as a luteal phase of gestation. Moreover, further menstrual cycles are suppressed throughout the pregnancy.

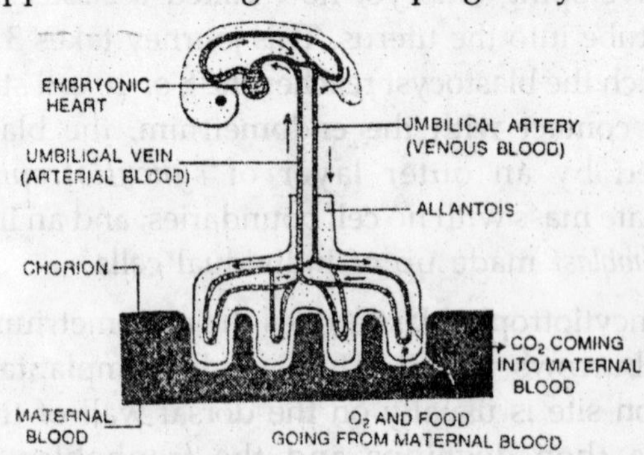

The embryonic blood circulation and the placenta.

In the human female the corpus luteum persists for about the first 12 weeks of pregnancy. During this luteal phase, progesterone output increases and the uterine lining becomes even more glandular and vasculansed. The early embryo, embedded and held fast within the wall of the uterus soon comes to consist of two groups of cells. The central group of cells forms the embryo proper. Surrounding this central group are the cells which form the so-called *extraembryonic membranes* (chorion, amnion, allantois and yolk sac). In the human female the chorion, as in the reptiles and birds, forms as an outer enclosure around the other membranes and the embryo being in direct contact with the tissue of the uterus.

In one region the chorion develops numerous finger-like outgrowths which branch extensively and erode paths through the spongy uterine wall. In this manner the tissues of the chorion and the uterus become firmly attached to each other. These interfingering and interlacing tissues are known as the *placenta.* When fully developed, the placenta functions both as a mechanical and a metabolic connection between the embryo and the female body. However, the embryonic and maternal bloods do not mix, being separated by the chorionic and allantoic membranes.

The placenta also functions as an endocrine organ. As it grows and develops, it manufactures slowly increasing amounts of estrogen and progesterone. In fact the progesterone output eventually becomes far greater than that of the corpus luteum. The corpus luteum actually degenerates and its hormone secretion subsides as the 12th week of gestation approaches. From that time on, the placenta provides the main hormonal control of pregnancy and through its progesterone output maintains its own existence. Thus, at about the 12th week, the luteal phase of gestation changes over to the placental phase.

(b) *Placental Phase of Gestation:* Because of the change in hormonal control, the 12th week is rather critical period of the pregnancy. If the corpus luteum should degenerate a little too soon and if the placenta should reach full development a little late, then the amount of the available progesterone during this gap is likely to be inadequate. In the absence of progesterone, the placental tissue could not be maintained. Just as at the end of a menstrual cycle, the uterine lining would disintegrate and the embryo, no longer anchored securely, would be aborted. Miscarriages occur frequently near the end of the 3rd month of the pregnancy. Such mishaps due to hormones deficiency can be prevented by injecting some progesterone into the pregnant female. The placental phase continues up to about 9 months during which all the bodily systems are formed.

Parturition: In correspondence with the growth and development of the foetus the amnion gradually enlarges and the uterus stretches. Under the stimulus of the increasing quantities of sex hormones from the placenta, the mammary glands enlarge markedly during the last months of pregnancy. The process of birth normally starts when the chorion and the amnion rupture, escaping the amniotic fluid to the outside. Once labour is started, the uterine contractions dilate the cervix, and this dilation in turn sets up signals in the afferent nerves that increase *oxytocin* secretion. Oxytocin increases uterine contractions in two ways: *(a)* it acts directly on uterine smooth muscle cells to make them contract, and *(b)* it stimulates the formation of prostaglandins in the decidua, which further enhance the oxytocin-induced contractions. During labour, spinal reflexes and voluntary contractions of the abodominal muscles ("bearing down") also aid in delivery.

Even though the source of progesterone is removed with the expulsion of the placenta, normal menstrual cycles in human female are generally not resumed as long as the nursing continues. The manufacture of lactogenic hormones by the adenohypophysis apparently inhibits FSH production. But once the child is weaned, FSH is formed again in quantity and a new ovarian follicle begins to mature.

Endocrine Regulation of Reproduction: Hormones control the development of sexual structures and the time of reproductive events. In the vertebrates the adenohypophysis, through its gonadotropic hormones, links the neurosecretory centres of the brain to the endocrine tissues of the gonads. Thus, neurosecretory centres, gonadotropins and gonadal hormones form a delicately balanced system for the timing of reproduction, the development of the gonads, and the differentiation of the secondary sexual characters. These various hormones together with the nervous system regulate the behaviour associated with the reproduction.

The endocrine tissues (androgen-producing tissue, and estrogen and progesterone-producing tissue) are found in the gonads of all vertebrates. From the most primitive to the most advanced forms they produce a series of steroid hormones whose synthesis is pituitary-regulated. This concert of hormones from the pituitary and the gonads performs together to regulate the reproductive process, and, while the pituitary gonadotropins stimulate the gonadal tissues, the gonadal steroids through a feedback mechanism serve to control the activity of the gonadotropic cells of the pituitary.

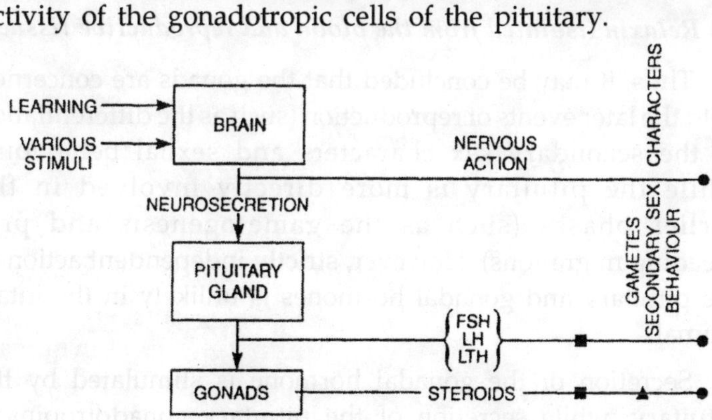

The control of reproduction in the vertebrates.

In brief, the endocrine regulation of the vertebrate reproduction is brought about by the following hormones.

(a) Gonadotropins

1. Follicle-stimulating hormone (FSH)
2. Luteinising hormone (LH)
3. Lactogenic hormone (prolactin)
4. Prolactin-like substances responsible for luteotropic action in rodents, stimulation of pigeon's crop glands for the production of pigeon's "milk", etc.

(b) Gonadal steroids

1. Androgens
2. Estrogens
3. Progesterone

(c) Placental hormones

1. Placental gonadotropins (Human chorionic gonadotropin HCG, Pregnant mare serum gonadotropin PMSG)
2. Chorionic thyrotropin
3. Chorionic growth hormones

(d) Relaxin (isolated from the blood and reproductive tissues)

Thus, it may be concluded that the gonads are concerned with the later events of reproduction (such as the differentiation of the secondary sex characters and sexual behaviour) while the pituitary is more directly involved in the earlier phases (such as the gametogenesis and pre-breeding migrations). However, strictly independent action of the pituitary and gonadal hormones is unlikely in the intact animal.

Secretion of the gonadal hormone is stimulated by the pituitary, while secretion of the pituitary gonadotropins is inhibited by the gonadal hormones. Within this broad framework of reproductive controls, interrelationships and their actions on the specific target organs are almost as variable as the animals themselves.

Contraception: Once conception has occurred, abortion can be produced by progesterone antagonists such as the synthetic steroid *mifepristone* (RU-486). In some countries mifepristone combined with a prostaglandin is used to produce selective abortions.

These days several methods are used to prevent conception. *Vasectomy* (bilatral ligation of the vas deferens) is considered to be a relatively safe and convenient procedure of male contraception. *Tubal ligation* is a procedure for female contraception. Intra uterine implantation of metal or plastic pieces (Copper 7, Loop D), which are called *intrauterine devices* (IUDs), also act as effective contraceptive method. Although the mechanism of action of IUDs is still debated, there is evidence that at least those containing copper exert a spermicidal action. However, the usefulness of IUDs is limited by their tendency to cause intrauterine infections.

Oral contraceptives are made of small as well as large doses of estrogen and progestin (> 50 μg estrogen and progestin, and < 50 μg estrogen and progestin). Progestin alone can also be used for contraception, although it is more effective when combined with estrogens. An orally active estrogen such as *ethinyl estradial* is often combined with synthetic progestin such as *norethindrone.*

The pills are administered for 21 days, then withdrawn for 5 to 6 days to permit menstrual flow and started again. Women undergoing long term treatment with relatively large doses of estrogen do not ovulate, probably because they have low FSH levels and multiple irregular bursts of LH secretion rather than a single midcycle peak. Women treated with similar doses of estogen plus a progestin also do not ovulate because the secretion of both gonadotropins is suppresed. Besides, progestin also makes the cervical mucous thick and unfavourable to sperm migration, and it may also interfere with implantation.

Among other methods of contraception are use of diaphragm, condom and spermicides, and withdrawal and rhythm.

Women are sometimes given large doses of estrogens for 4 to 6 days to prevent conception after coitus during the fertile period (postcoital or "morning after" contraception). For emergency contraception, emergency contraceptive pills are used.

These are basically regular oral contraceptive pills. The dosage is simple, 2 pills as soon as possible and 2 pills 12 hours later. This therapy, known as the Yuzpe regimen, is named after the Canadian doctor who discovered it, and is widely used in developed countries.

5

System of Mating

The sexual selection is based upon the struggle for mating. Sexual selection implies presumptive choice on her part of the most desirable male to be the father of her progeny. Biological fitness provides an ability to contribute to gene pool of the next generation and subsequent generation to reproduce. An animal that successfully copes with its problems of obtaining food, water and shelter but fails to reproduce still may have fitness of zero. Reproductive behaviour is of obvious importance to fitness and lies very close to many definition of species.

Reproductive behaviour possesses a set of related questions about mating and rearing of offspring. Before conception, there may be a struggle among males and a characteristic sequence of behaviour pattern in which the sexes court each other. The term courtship is used by the ethologist refers to behavioural interaction for the fertilization. In some species, such as stickle back it lasts a few minutes while in others may last for month but in some species it does not even exist.

Courtship ensures that animals mate with other individual of the same species, sex and condition.

Choosing a Member of the Right Species: Animals do not mate with different species because it is adaptive in animals. Mating between members of different species are very rare. It is proved by several examples, in a sample of 725 copulating pair of two species of cicada, which are distinguishable but similar in appearance of human eye, only seven contained members of two species while rest were mating of the same species. Hybrid individuals are inferior or weaker to those produced by mating of members of the same species because they are sterile. The best-known example of a sterile hybrid is the mule. Natural selection will favour animals that produce normal healthy offspring by choosing to mate with members of their own species. There have been many investigation factors for the enshrinement that they mate with right species. In case of crickets there are approximately 3000 species of cricket in the world. Male cricket, produce a song for attracting the female of same species. The songs are produced by males when they rub their wings together and are heard by the female through the ears on the front legs.

The all importance of song in the species recognition of cricket can be shown by another experiment. Female cricket can be trained to walk along a "y-maze" on which they have to turn to the left or right. If for example the loudspeaker on the left were playing the song of the female's own species, while the one on the right played the song of some other species, the female turns to the left, the side of the song can be reversed to control for any bias the female may have for turning in one direction or the other.

Need of Sexual Selection

Individual can be successful by the quality of probable offspring. Thus, it becomes important who its mate will be. Darwin (1871) introduced the concept of sexual selection. Sexual selection can be divided into two types: one in which members of one sex choose certain mates of the other sex (intersexual selection), and a second in which individual of one sex compete among themselves for access to the other sex (intrasexual selection)

Intersexual Selection: In intersexual selection, individual of one sex shows that they are worthy of an investment so the member of other sex choose among them. Fisher (1958) used birds as an example. Bowerbirds are unique in that they build a structure whose only function is to attract mates. The male satin bowerbird (Ptilonorhynchus violaceus) of Australia and New Guinea station itself alone in the forest, clears a space weaves the bower from twigs, and decorates it with brightly coloured objects. Gilliard (1963, 1969) reasoned that bowers function in place of showy plumage, and noted that those species with the fanciest bowers had drabest plumage.

Intrasexual Selection: It is a competition (usually males) within one sex. The winner gaining access to the opposite sex. This competition may take place in several animals such as deer (family Cervidae) and antelope (family Bovidae). The winner gain dominance and do most of the mating.

Females may incite competition among males and thus may gain some control over the choice of mate. For example, female elephant seals (Mirounga angusiirostris) vocalised loudly whenever a male attempt to copulate. This behaviour attracts other males and tests the dominance of male attempting to mate. In some other cases a male will sometimes attack a mated pair, displacing the previous male and mating with the female.

Among insects such as moths and butterflies, two kind of sperms are produced. One is eupyrene that fertilizes the eggs and the second type apyrene contain no nuclear material. Silberglud, Shepherd and Dickinson (1984) suggest that apyrene are the result of sperm competition, possibly displacing eupyrene sperm from first male or delaying remating by the female. Following conception, male-male competition may take a different form. In mice, the "Bruce effect" operate early in pregnancy, a strange male causes the female to abort and become receptive. Among langeur monkeys (Presbytis entellus) strange male may take over a group, driving out the resident male. The new male may then kill the young sired by the previous male (Hardy 1977). Females who have lost their young soon become sexually receptive.

Timing or Seasonally: Animals breed in a particular time or season. So these animals are known as seasonal breeders. In lower animals, timing and seasonality is less understood. But in higher animals, it is very complicated affair. In fishes, birds and mammals of the northern temperate zone, reproduction begins in spring. The first phase is migration and is done by all individual at approximately the same time. In some cases, the timing process is not very accurate. There may be a considerable difference between the male and female of a pair. It has been found in pigeons and in other animals. When a male and a female are kept in adjoining cages so that they can see and even touch each other, but are prevented from copulating the persistent courtship of the male will finally induce the female to lay eggs. These eggs are infertile.

Definition of Sex Ratio

Suppose there are 20 females for every male in a population. Every male has 20 times the expected reproductive success of a female and therefore a parent whose children are exclusively sons can expect to have almost 20 times the number of grand children produced by a parent with mainly female offspring. A female biassed sex ratio is therefore not evolutionarily stable because a gene which causes parents to bias the sex ratio of their offspring towards males would rapidly spread a greater proportion of males than the initial 1 in 20. But now imagine the converse, if males are 20 times as common as females, a parent producing only daughter will be at an advantage. Since one sperm fertilizes each egg. Only one in every 20 males can contribute genes to any individual offspring and females therefore have twenty times the average reproductive success of a male. So a male biassed sex ratio is not stable either. The conclusion is that the rare sex always has an advantage and parent which concentrate on producing offspring of the rare sex will therefore be favoured by selection. Only when the sex ratio is exactly 1: 1 will the expected success of a male and a female be equal and the populations label. This means that parents should have invested equally in sons and daughters.

(a) *Local Mate Competition:* We can understand it by following examples. Suppose there are two sons and they have only one chance to mate and that they compete for the same female. Only one of them can be successful in mating so according to their mother's point of view one of them is wasted. Therefore, the mother should bias her ratio of investment towards daughters.

The exact degree of bias predicted by Fishers theory depends on the degree of local mate competition. Extreme competition is to be expected in species with limited powers of dispersal, and therefore, it tends to be associated with inbreeding. The best sex ratio in this instance is to produce just enough sons to fertilize the daughters, since any other mates will be wasted. The crucial difference between this and the earlier argument for a 1: 1 sex ratio is that here the ratio of males to female in the rest of the population does not matter.

(b) *Local Resource Competition or Enhancement:* Anne Clark found that the South African prosimian Galago crassiccudatus has a male biassed investment ratio among its offspring. As with most mammals, female Galago disperse less far than males, and often end up competing both with their mother and with each other for rich sources of food such as gum and fruit trees in the mother's home range. This local resources competition among females reduces their value as offspring in the extreme case only one daughter might be able to survive on the food available near home and so investment in other daughters would be wasted.

(c) *Maternal Condition:* Male red deer always compete for the females by prolonged roaring and antler wrestling contest. In these contests, it is an advantage for a male to be big and size depends among other things on how well-fed the male was as a youngster, which in turn depends on his mother's ability to compete for good source of food and hence produce a plentiful supply of milk. Now if a mother knew that her sons would be highly successful harem holders, it would pay her to invest heavily in sons would not grow up to be big and strong would be better to have daughters. Since a daughter's future reproductive success does not depend so much on her mother's milk. Exactly, this pattern has been found in red deer.

Ardent Males: During mates, male often fight and ritualised and have evolved weapons for fighting. Male may dispute over direct access to females. The most intense fights in many species occur when females are ready to be fertilized and once a male finds a female he often guards.

In some invertebrates, the male cements up the female's genital opening after copulation to prevent other males from fertilizing her. The male of moniliformes dubius parasitic acanthocephalan worm in the intestine of rats, produces a chastity belt of this kind but in addition to sealing up the female after copulation. The male sometimes copulate with rival males and applies cement to their genital region to prevent them from mating again. A male xylocoris may inject his sperm into a rival male. The sperms then swim and wait to be passed on to a female.

The other example of the acrane method of male-male competition found among invertebrate is the use of anti-aphrodisiac smells. Larry Gulbert noticed that female Heliconius erato butterflies always smell peculiar after they have mated. He was able to show experimentally that the scent does not come from the female herself but is deposited by the male at the end of mating. Gilbert also found that the scent discourages other males from mating with the female.

Reluctant Females: Females in the great majority of species are the chief resources of zygote. Not very surprisingly, females are the whole chooser than males during courtship, by which female get or obtain good genes for their offspring.

(a) *Good Resources and Parental Ability:* For the betterment of the young ones, male often defend breeding territories containing resources which play a crucial role in the survival of a female's egg or young. For example, male North American bullfrogs defend territories in ponds and small lakes where female come to lay their eggs. Some territories are much better for survival of eggs than other and these are the ones which females prefer. The bullfrogs also show that female choice and male-male competition may go hand in hand.

Food is a resource which often limits a female's capacity to make eggs and during courtship females may choose whether or not to mate with a male on the basis of the ability to provide food. In some birds and insects, for example, males may provide food for the females during courtship which makes a significant contribution to her eggs. Female hanging flies (Hylobitacus apicalis) will mate with a male only if he provides a large insect for her to eat during copulation and the more eggs he fertilizes. In birds the male usually helps to feed the young and courtship feeding may play the additional role of indicating to the female how good the male is in bringing food for the young.

Significance of Good Gene

Good gene are ones which increase the ability of her offspring to survive, compete and reproduce. Linda partridge experimented on female fruitflies (Drosophila) and allowed them to mate freely with a population of males or forced each female to mate with a randomly choosen partner. The offspring of the "choice" and "no choice" females were then tested for their competitive ability by rearing the larvae in bottles with a fixed number of standard competitors. Partridge concluded that the offspring of the choice group did slightly but consistently better than those of no choice females in the larvae competition experiments. This experiment suggests that females are able to increase the survival of their offspring by choosing good genes in their mates but it must be born in mind that the results could also be in part explained by intrasexual competition in the choice experiment the males that mated may have superior competition against other male.

Male Investment: In many birds, some amphibians and arthropods both male and female invest about equally in the eggs or young by feeding guarding or brooding. Sometimes the usual sex roles are completely reversed so that males do the investing and female competing. The idea about sexual conflict and sexual selection can still be applied in modified form to species with equal or primarily male investment. When both sexes care equally for the offspring, for example, courtship may involve assessment and choice by males as well as by females. Males of species with internal fertilization can never be absolutely sure that they have fathered the children of their mate.

Sexual Clash

Recall the view of the origin of anisogamy as the primaeval example of sexual conflict. The conflict was one about mating decision. Today, the conflict is not only with respect to mating decision but also in the contexts of parental investment, multiple mating and infanticide.

(a) *Mating Decision:* As we already know that female tends to be more choosier than males. An extreme manifestation of this conflict is enforced copulation as exemplified by scorpionflies (Panorpa spp). Male scorpionflies usually acquire a mate by presenting her with a nuptial gift in the form of a special salivary secretion or a dead insect. The females feed on the gift during copulation and turns the food into eggs.

However, sometimes a male enforces copulation: he grasps her with a special abdominal organ enforced copulation appears to be a case of sexual conflict. The female loses because she obtains no food while the male benefits because he avoids the risky business of finding a nuptial gift.

(b) *Parental Investment:* It is sufficient to note that in species with investment beyond the gamete stage, each sex might be expected to exploit the other by reducing its own share in the investment. The outcome of this sexual conflict may depend on practical considerations such as which sex is the first to be in a position to desert the other which has the possibility of deserting the female immediately after fertilization and leaving her with the eggs or young to care for it.

(c) *Infanticide:* Male lion may slaughter the cubs in a pride shortly after they take over as group leaders. This behaviour increases male reproductive success and clearly decrease female success. This seems to be a case of sexual conflict in which the males have won but it is perhaps surprising that females have not evolved counter adaptation. For example, they killed their young ones to recoup as much as possible of their losses.

(d) *Multiple Mating:* When a female mates with more than one male termed as multiple mating. Multiple mating is always advantageous to the male. For example, when two dungflies males struggle for a female the female is sometimes drowned in cow-dung due to the fighting males on top of her.

Mating System: Despite the argument based upon relative parental investment by the sexes in animals polygamy should be the normal mating system, whereas monogamy is common in some groups for example in 90 per cent of birds. In many birds parental investment is same for male as well as females. Young altricial birds in nests are very vulnerable to predation and survive best if they can develop rapidly. If the female alone can seldom rear two young, it is advantageous for males to help with parental care.

Emlen and Oring (1977) summarise the essential conditions for the occurrence of polygamy in terms of the relative economics of defending one or several males. This depends upon the spatial distribution of resources, the sex ratio and the time at which mates become available. If female exhibit synchrony in their sexual receptivity, polygamy is more difficult. An intermediate between polygamy and polyandry in which female lay a second batch of eggs which the male incubates or guards but where the male also mate with other females was rapid multiple clutch polygamy. This is common in fishes and rare in birds. Polyandry has not been reported in mammals (Eisenberg 1966) but Jenni (1974) describes it in several species in birds.

The females of American Jacana (Jacanaspinosa) defend large territories in march areas which include the smaller territories of several males. The males perform all incubation and parental care but they defend their territories against other males. Emlen and Oring also described harem polyandry and female access polyandry in addition to resource defend polyandry.

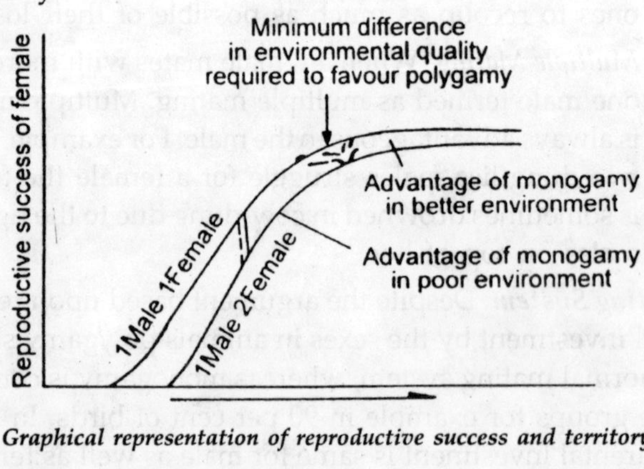

Graphical representation of reproductive success and territory

The Struggle of Males and Females: Due to the reason that the males reproductive success is probably limited by the number of females they can attract and defend from other males rather than their sperm supply natural selection will favour any property in a male that enables him to mate with more females. The most obvious form of competition for females is straight-forward fighting. In toads for instance, the males sit on the back of the females for a few days before the female lays her eggs. Other males try to dislodge the sitting males from the females by pulling them off. In experiments larger males are much better at dislodging smaller males from the females than vice versa.

There is an advantage in being large in males, and large males are more likely to mate with females than are small males.

In species that fight, a male can increase his effective strength by forming a coalition with another male, and ganging up on competitors. In lions, a group of two or three male lions try to take over harems of females by forcibly erecting the existing male owners. Bigger coalitions of males are more successful in taking over harems.

However, physical fighting is not the only means by which males compete with each other. When a coalition of male lions successfully takes over a harem of females, the jut thing the males do is to kill all the young lion cubs in the pride. The cubs of course, were fathered by the previous males, and are no loss to the new owners. There is even a gain to them. When a lioness is locating for her cub she will not produce another cub but if her cub is killed she soon becomes ready to reproduce again.

Courtship: Courtship phenomenon bring together two animals of different sexes of the same species under conditions where mating is likely to occur and to be successful. The initial problem is to locate a potential mate and obviously the conspecific must be of the opposite sex if reproduction is to be successful. Courtship generally entails a complex sequence of interacting signals, that function to ensure that an animal mates with an appropriate partner. Timing is an important part of successful reproduction; both male and female must be in an appropriate physiological condition for mating, although mating system varies greatly from species to species.

In arthropods, initial detection occurs via olfaction for example silk moths, Bombyx mori. These are noted for the extreme sensitivity of the male to a sex attractant, bombykol, produced by the female. There are several other combination of behavioural patterns like wing vibrations, leg vibration, wing seamaphoring, circling and licking.

In fishes various species display diverse courtship patterns. In the tropical aquarium fishes, behaviour has been studied very carefully. Blackchin mouthbreeders, (Tilapia melanotheron), display four pure courtship patterns — patterns that occur in virtually no other contexts. These are the nod, *i.e.*, rapid tilting down; quiver, *i.e.*, a pattern of head-shaking; nib, *i.e.* biting the substrate and skim, *i.e.*, remaining motionless near the nest (Barlow and Green 1970). In the courtship of guppies the most prominent display is the adoption of "sigmoid' or S-shaped posture by the male. In amphibians the male bullfrogs adopt territories from where they emit their familiar loud choruses to attract the females.

In reptiles the crocodiles also emit loud roars. Courtship in most species of snakes is based on tactile stimulation of the female and olfactory stimulation of the male. The so-called dramatic courtship dances in which two snakes become closely intertwined have usually proved to be male-male interactions which may be aggressive in nature (Porter 1972).

In birds there are many dramatic examples of courtship behaviour. The complex displays of grubes, gulls, ducks, herons and other birds have been favoured subjects of study by ethologists. In green herons male set up territories. The full forward display functions to drive away other males from a resident male's territory. The calls of the male attract the female.

Initially the females are repelled with full forward displays, but later on they persist and courtship is eventually initiated with the appearance of the snap display and the stretch display. The two birds fly about the territory after pairing and occasionally exhibit the more intense flap-flight display. According of Meyerriecks (1960) until copulation displays patterns become more contact oriented.

In most of the birds, their song functions to repel males from territories and to attract females.

Among mammals Ewer (1968) surveyed some courtship patterns and according to him olfaction plays a large role in the regulation of courtship in most of the species. Patterns of anogenital investigations and urine testing are common. Female mammals often solicit mountings, sometimes by approaching a male, nuzzling or licking him and often by darting from him.

In bottle-nosed dolphins, courtship entails vocalisations, mouthing of the partner's genitalia, rubbing of bodies, displaying of the white underside, chasing, head butting, leaping. (Puente and Dewsbury, 1976)

Mating: Actual mating occurs as a result of consummation of courtship activity where the fertilization of ova is accomplished.

In most of the species of arthropods, sperm are 'packaged' in a spermatophore, which is a kind of bag or sac containing the sperm. Many aquatic species have external fertilization for, *e.g.* Limulus, it hooks itself on behind the female and remains there until the eggs are released. Most of the terrestrial forms exhibit internal fertilization.

In fishes fertilization may be either external or internal. In anabantid fish like bettas and gouramеs, a bubble nest is prepared and spawning generally occurs beneath it. Eggs and sperms are released simultaneously and egg are fertilized as they float up to the nest.

In amphibians, many species of salamanders exhibit internal fertilization through use of a spermatophore, although parthenogenesis and external fertilization are also of common occurrence. Most frogs and toads have external fertilization, eggs are fertilized as they are laid. Exceptions include tailed frogs, which have an intromittent organ, and African toads in which cloacal glands of the male and female are brought together. Rabb in 1973 stated that if male clasps another male or in clasping a female that has finished spawning, a release call by the recepient results in release by the clasping male.

In reptiles the fertilization is internal through the use of intromittent organs.

In birds the copulations generally show little variabilities. It most of the species penis is lacking, thus the transference of sperm occurs from cloacal glands of the male to that of the female.

In mammals a lot of research on copulation is being done using laboratory rats Rattus norvegicus. Its copulatory behaviour is described in detail. Whenever a receptive female is introduced into a cage containing a vigorous male, a predictable sequence of events occurs. After initial courtship behaviour like sniffing, chasing, etc., the male persues the running female. As the male mounts the female from behind, the female adopts a stereotyped posture, called "Lordosis" The head and hindquarters get elevated as a result back is produced into a concave arch, tail remains deflected to the side. Two kinds of copulations occur.

In some called 'intromissions' the male inserts its penis into the vagina of female for a period of 1/4 second before the male displays a rapid dismount. Although prior to insertion male may exhibit extravaginal pelvic thrusting, but there is no intravaginal thrusting. After about ten such short intromissions, the male displays 'ejaculation'. During ejaculation the male shows a kind of convulsive thrusting and clasps the female for a few seconds before he dismounts. A complete group of intromissions ending with an ejaculation is known as "series". Without prior intromissions there is no ejaculation and sperm transfer ejaculation is followed by a temporary cessation of sexual activity.

After about five minutes, activity resumes and a second series occurs. Rats normally completes about seven ejaculatory series before they may be considered satiated, according to a criterion of half an hour without intromission (Beach and Jordan, 1956).

6

Suspension and Buffer

A mixture composed of particles that are dispersed in another medium may be one of the three categories: *solution, suspension* or *colloid*. In a solution, the individual molecules or ions of a substance, the *solute,* are small enough to dissolve in a liquid medium, the solvent. These molecules or ions are so small (less than 1×10^{-7}) that they never settle out under the influence of gravity. The mixture is homogeneous, *i. e.,* perfectly uniform in appearance. Molecules or ions of the solute in a solution cannot be filtered out. If a solution is heated vigorously the solvent evaporates and the solute particles separate out as crystals. Therefore, solution is also sometimes referred to as a *crystalloid* (Thomas Graham, 1861).

Sugar dissolved in water provides the best example of a true solution. When a solid is dissolved in water upto the limiting concentration, it is called a *saturated solution.*

Definition of Suspension

A mixture like that of sand and water is called a suspension. The sand particles (crystals of SiO_2) are so large (more than

2×10^{-5}) that after a period of time, they settle by gravity to the bottom of the container. It is easy to separate sand and water by filtration.

Colloids: The term *colloid* is derived from Greek *koli,* glue and *eidos,* appearance. Colloids are suspensions of particles that are larger than those in true solution but still too small to settle out by gravity. They are of intermediate size, from about 1/1,000,000 to 1/10,000 mm (1×10^{-7} to 2×10^{-5}) in diameter. They cannot be filtered out by ordinary filters. The white of egg is a familiar colloid.

Types of Colloids: There are eight general types of colloidal systems:

(A) A colloidal *gas* dispersed in

(i) a solid, *e.g.,* cheese: colloidal air in fat-protein.

(ii) a liquid, *e.g.,* Campa Cola, Thums Up, etc.

(B) A colloidal *liquid* dispersed in

(iii) a liquid *(emulsion), e.g.,* milk: colloidal fat in water.

(iv) a solid *(gel), e.g.,* gelatin: colloidal water in protein.

(v) a gas *(aerosol), e.g* , fog: colloidal water in air.

(C) A colloidal *solid* dispersed in

(vi) a liquid *(sol), e.g.,* milk of magnesia: colloidal $Mg(OH)_2$ in water, and starch dispersed in water.

(vii) a solid, *e.g.,* colloidal gold in glass (rubi glass).

(viii) a gas *(aerosol), e.g.,* cigarette smoke: colloidal ash in air.

Nature of Colloids: A colloidal system consists of two heterogeneous components: a *dispersed phase* and a *dispersion phase* or medium.

(i) *Dispersed or discontinuous phase* comprises macromolecular solids like proteins and nucleic acids; and liquids such as oily fats. The dispersed particles of the discontinuous phase are called *micelles.*

(ii) *Dispersion or continuous phase* is the medium in which insoluble materials of colloidal size are dispersed. Dispersion medium can be solid, liquid or gas. In protoplasmic colloidal systems, this medium is water. The particles of the dispersion phase are called the *intermicellar particles.*

Properties of Colloids:

(1) *Electrical Properties:* The particles of a colloidal solution carry similar electric charges which may be positive or negative (colloidal ions). These charges make the particles repel one another and this keeps them suspended. Every ion in the dispersion medium has an opposite and equal charge so that the system as a whole is electrically neutral.

(2) *Brownian Movement:* The dispersed particles of a colloidal system are constantly moving in random zig-zag paths. This haphazard motion of the dispersed particles in the dispersion medium is called the Brownian movement (after Robert Brown, 1828). The smaller the particle size the more violent the Brownian movement. This type of movement is due to the bombardment of the colloidal particles by molecules of dispersion medium.

(3) *Tyndall Phenomenon:* Light shining through a colloidal mixture gives it a cloudy or opalescent appearance as opposed to a true solution which is optically clear. This is called the Tyndall phenomenon. It results from the light scattering properties of colloidal dimensions. When a beam of light is passed through a colloidal solution, the path of the beam of light can be observed from the side because the dispersed particles are big enough to scatter the light. This effect is used to distinguish colloids from true solutions.

(4) *Sol and Gel:* A colloid may remain in two states, as a liquid or as a solid. The colloidal solution or *sol* is a

system of solid particles of colloidal dimensions dispersed in a solvent medium. The *sols* are divided into two classes:

(i) *Lyophilic (Liquid Loving):* These are the reversible colloids which after coagulation can be again converted into colloidal state by adding the solvent. Examples are glue, starch and gelatine. Formerly, the sols in which water is the dispersed medium were called hydrophilic (water loving), but now-a-days the more general term "lyophilic" is used because there are several colloidal solutions in which dispersion media are other than water.

(ii) *Lyophobic (Liquid Fearing):* There are the irreversible colloids which can readily be precipitated from the solution. They do not regain their colloidal state even if the solvent is added. For example, the colloidal solution of gold, once coagulated, cannot be reconverted into colloidal state by dipping it into the solvent.

If liquid particles are dispersed in a solid medium the colloidal system is known as a *gel* which is like a quasi-solid, flexible gelly, as in gelatin and fruit jellies. The semi-solid structure of a gel is due to the solid containing a liquid arranged in an intricate meshwork extending throughout the system. Many long-chain proteins and high molecular nucleic acids are commonly found as gels.

The cytoplasm of a living cell is normally in the gel state but it undergoes continuous, dynamic, localised and reversible solgel transformations, also called *phase reversals*. In living materials, a region remains more or less permanently in a gel state, like skin, or in a sol state, like blood plasma. However, the latter becomes a gel on clotting. Gels maintain definite forms and shapes despite high water contents.

Increased temperature may convert a gel into a sol. This change is called *solution.* Conversely, lower temperatures promote conversion of sols to gels, a process known as *gelation.* Many other physical and chemical influences, such as low or high pH or pressure, affect solgel transformations.

(5) *Dialysis:* If a mixture of crystalloids and a colloid is placed on one side of parchment membrane facing water on the other side, the colloid particles would be retained on the same side, whereas crystalloids and a great deal of dispersion medium would pass out through the parchment membrane to the other side. This phenomenon is called "dialysis" which is employed in "artificial kidney" to remove urea and other unwanted small metabolic wastes from plasma proteins.

(6) *Ageing:* Colloids undergo ageing. The particles in a fresh young colloidal system are surrounded by layers of *bound water.* Colloid particles *imbibe* a good amount of water. The molecules of water are *absorbed* to the surface of the colloidal particles by electric attraction. Adsorption is a peculiar form of combination in which substances adhere together by surface contact. It is not a true chemical reaction. Water in the gel, therefore, does not "run out" through its lattice work. As time passes, binding capacity of the particles diminishes, and some of the water runs out. The colloid contracts and undergoes gelation gradually. Examples are exudation of water from long-standing milk curd and custard.

(7) *Flocculation or Coagulation:* In colloidal solution all particles have similar charge and they remain separate due to forces of repulsion. When the charge on the particles is neutralised by the addition of ions or micelles of opposite charge, the particles undergo collision because of Brownian movement and *agglomeration* of the individual particles takes place. It results in the formation of larger particles which settle down out of

the system. This phenomenon, is called flocculation or precipitation.

(8) *Formation of Boundary Membranes:* An interfacial boundary membrane is formed between a colloidal system and a different medium such as air, water or solid surfaces, *e.g.*, the 'skins' on boiled milk, pudding, custard, etc.

Biological Significance of Colloids: Colloids are of immense biological value. Some examples are as follows:

(1) *Cell protoplasm* in every cell exists in a colloidal state.

(2) *Blood plasma, Lymph* and *milk* are all colloidal emulsions.

(3) *Blood clotting* is a colloidal phenomenon in which plasma (sol) is converted into a cloth (gel).

(4) *Imbibition of Water:* Since colloid particles imbibe water, a good amount of water remains stored in the body.

(5) *Interfacial Reaction:* Colloids are dispersed in the form of minute particles, hence they provide a very large surface area for various reactions to occur, such as enzyme action, adsorption, surface tension, etc.

Hydrogen Ion Concentration (ph) (Potentia Hydrogen): The number of hydrogen ions present in a solution is a measure of the acidity of the solution.' The maintenance of the normal hydrogen ion concentration within the body is an important factor in the environment of cells.

A standard scale for the measurement of the hydrogen ion concentration in solution has been developed. All acids do not ionise completely when dissolved in water, that is, all the molecules of acid in solution do not ionise and exist in the solution as electrically charged particles. The hydrogen ion concentration is a measure, therefore, of the amount of *dissociated acid* rather than of the amount of acid present. Strong acids dissociate more freely than weak acids, while carbonic acid dissociates much less freely into H^+ and Cl^-, while carbonic acid dissociates much less freely into H^+ and CO_3^-. The number of *free hydrogen ions* in a solution is a *measure of its acidity* rather than an indication of the type of molecule from which the hydrogen ions originated.

The alkalinity of a solution is dependent upon the number of hydroxyl ions (OH^-) present. Water is a neutral solution because every molecule contains one hydrogen ion and one hydroxyl radical. For every molecule of water (H.OH) which dissociates, one hydrogen ion (H^+) and one hydroxyl ion (OH^-) is formed, each one neutralising the other.

The scale for measurement of pH was developed taking water as the standard. It was found by experiment that 1 mole in 550,000,000 moles of water ionises into a hydrogen ion and a hydroxyl ion. This is the same proportion as 1 gram hydrogen ion in 10,000,000 litres of water. Therefore, 1 litre of water contains $\frac{1}{10,000,000}$ of a gram of hydrogen ion. In order to make these figures more manageable this can be written:

1 litre of water contains $\frac{1}{10^7}$ g hydrogen ion.

$(10 = 10^1 \; ; \; 100 = 10^2 \; ; \; 10,000 = 10^4)$.

The fraction $\frac{1}{10^7}$ may be written 10^{-7} the negative power indicates that this is a fraction, $\frac{1}{10,000,000}$ may be written $\frac{1}{10^7}$ or 10^{-7}.

Later, for every day use, only the power' figure was used and the symbol pH placed before it.

Thus a neutral solution such as water, where the number of hydrogen ions is balanced by the same number of hydroxyl ions, the pH = 7. The range of this scale is from 0 to 14. If the pH is 0 it would mean that 1 litre of water contained 1/1 = 1 gram of hydrogen ion ; or at the other end of the scale if there were no hydrogen ion present it would be written $\frac{1}{10^{14}}$ or pH 14. It will be noted that a change of pH of 1 at any level in the scale means an increase or decrease by a factor of 10 hydrogen ion concentration because the pH scale is a logarithmic one.

The pH scale thus runs from 1 to 14, neutrality being at pH 7.0. Numbers lower than 7.0 denote an excess of hydrogen ions and an *acid* solution ; numbers *higher* than 7.0 denote an excess of hydroxyl ions and an *alkaline* solution.

pH Values of the Body Fluids: All body fluids have pH values which must be maintained within relatively narrow limits and within which the cells function normally. These pH values are not the same in every part of the body, for example, the pH values of the following secretions are:

Blood	7.4 to 7.5
Bile	7.8 to 8.6
Pancreatic juice	8.0
Saliva	6.4 to 6.8
Urine	5.5 to 7.0
Gastric juice	1.5 to 1.8

It can be seen that there is a wide variation of pH values within the body. The individual pH in an organ is produced by it's secretion of acids or alkalies which will establish the optimum level. The highly acid pH of the gastric juice is maintained by the secretion into the stomach of hydrochloric acid produced by chemical changes which take place in parietal cells in the wall of the stomach. The low pH value in the stomach provides the environment best suited to the functioning of the enzyme pepsin which is present in gastric juice. Saliva has a pH of between 6.4 and 6.8, which is the optimum value for the action of ptyalin. Ptyalin action is inhibited when food containing it reaches the stomach and is mixed with highly acid gastric juice.

It will be noted that the blood has a pH value of between 7.4 and 7.5. Thus, therefore, it is the general pH level in the body which may be altered in an individual organ such as the stomach, to meet the needs of the special functions of that organ. The range of pH of the blood which is compatible with life is 7.0 to 7.8. The metabolic activity of the body cells produces certain acids and alkalies which tend to alter the pH of the tissue fluid and the blood. To maintain the pH within the normal range there are substances present in the blood which act as *buffers.*

Characteristics of Buffers

Buffers are chemical substances which are able to 'bind' free hydrogen ion or hydroxyl ion without change in pH. Three major buffer systems are found in animals, and these involve *phosphate, bicarbonate,* or *amino acids.* A certain amount of all three is normally present in the body and cellular fluids, and so the maintenance of a constant pH depends on a complex system.

(i) *Phosphate System:* A solution of sodium dehydrogen phosphate ionises in solution in the following manner :

$$NaH_2PO_4 \rightleftharpoons Na^+ + H_2PO_4^+ \rightleftharpoons H^+ + HPO_4^{2-} \rightleftharpoons H^+ + PO_4^{3-}$$

The final ionisation in this series is only half complete at a pH of nearly 13.0, and may be ignored from a physiological viewpoints. The predominant phosphate ions in solution at a physiological pH will therefore be $H_2PO_4^-$ and HPO_4^{2-}. Phosphate buffers are, in fact, usually made up as a mixture of the two salts, sodium dihydrogen phosphate NaH_2PO_4, and disodium hydrogen phosphate Na_2HPO_4. It is evident from the scheme of ionisation given above that the second of these two salts act as a weak acid, since it produces some hydrogen ions in solution, whereas the first salt represents the salt of phosphoric acid H_3PO_4 with the strong base NaOH. The mixture of these two salts will behave like the generalised buffer. Thus, if hydroxyl ions are added, water is formed:

$$NaH_2PO_4 + NaOH \rightarrow Na_2HPO_4 + H_2O$$

The addition of acid results in the formation of the basic salt:

$$Na_2HPO_4 + HCl \rightarrow NaCl + NaH_2PO_4$$

(ii) *Bicarbonate System:* The bicarbonate ion HCO_3^- can take up hydrogen ions from water to form carbonic acid:

$$HCO_3^- + H^+ \rightleftharpoons H_2CO_3$$

Although little carbonic acid is formed under normal conditions, the addition of alkali to this solution will cause it to react with the carbonic acid, remove it through the formation of water, and so facilitate the production of more carbonic acid:

$$H_2CO_3 + OH^- \rightarrow H_2O + HCO_3^-$$

The net effect is that the added hydroxyl ions have been removed from solution, with the formation of more bicarbonate ion. If acid is added, it will react with the bicarbonate ion to produce carbon dioxide and water:

$$H^+ + HCO_3^- \rightarrow H_2O + CO_2 \uparrow$$

Some of the carbon dioxide will go into solution and form carbonic acid, but since this is a weak acid, it's effect is negligible. Thus bicarbonate can act as a buffer because it forms a weak acid in solution, and it's action conforms to the generalised scheme previously described.

Sodium bicarbonate ($NaHCO_3$) is normally present in blood and acts as buffer to acid radicals like lactate, phosphate and sulphate which are stronger than carbonate.

(iii) *Amino Acids System:* Amino acids contain in their molecule both a basic $-NH_2$ group and an acidic -COOH group. They can be visualised as existing in the form of a neutral *zwitterion* in which a hydrogen atom can pass between the $-NH_2$ and -COOH groups, for example in the case of *glycine*, the molecule can be depicted as

$$\begin{array}{c} NH_3^+ \\ | \\ H-C-COO^- \\ | \\ H \end{array}$$

By the addition or subtraction of a hydrogen ion to or from the zwitterion, either the cation form or the anion form will be produced:

$$\underset{\text{cation form}}{H_3N^+-CH_2-COOH} \underset{}{\overset{+H^+}{\rightleftharpoons}} \underset{\text{zwitterion}}{H_3N^+-CH_2-COO^-} \underset{-H^+}{\rightleftharpoons} \underset{\text{anion form}}{H_2N-CH_2-COO^-}$$

Thus, when OH′ ions are added to the solution of amino acid, they take H^+ ions from it to form water, and the anion is produced. If H^+ ions are added, they are taken up by the zwitterion to produce the cation form. In practice, if NaOH was added, the salt $H_2N—CH_2—COONa$ would be formed, and the addition of HC1 would result in the formation of the hydrochloride $ClH—H_3N—CH_2—COOH$, but these substances would ionise in solution to some extent, to form their corresponding ions. *Haemoglobin* and *plasma proteins* act as buffers in a similar way.

Amino acids differ in the degree to which they will produce the cation or anion form. In other words, a solution of an amino acid is not neutral, but is either predominantly basic or acidic, depending on which form is present in greater quantity. For this reason, different amino acids may be used as buffers for different pH values, and a mixture of them possesses a wide buffer range.

The acids which are constantly added to the tissue fluids by the cells are first buffered and then disposed of in various ways. Lactic acid resulting from muscular contraction is partly converted back into glycogen by the liver. Phosphates ; sulphates and chlorides are excreted by the kidneys. Carbon dioxide, which is continually produced by all cells, is removed in the lungs which thus exert an accurate and rapid control over the acid-base balance.

The ability of the complex buffer systems of the blood to 'bind' hydrogen ions or neutralise acids is called the *alkali, reserve* of the blood. If the pH of the blood falls below 7.4 the reserve of alkali has been reduced by an increase in production of hydrogen ions by the cells, and the condition of *acidosis* exists. When the reverse situation pertains and the pH is raised above 7.5 the increased alkali produced uses up the *acid reserve* and a state of *alkalosis* exists.

Ionisation

Almost all chemical reactions of biological importance occur in aqueous solution. Compounds which give rise to ions in solution are called *electrolytes.* For example, when solid NaCl (an ionic compound) dissolves in water, Na^+ and Cl- ions leave the crystal and are available among the water molecules. This separation into free ions is termed *ionic dissociation.* For example,

$$\underset{\text{sodium chloride}}{NaCl} \xrightarrow{H_2O} \underset{\text{sodium ion}}{Na^+} + \underset{\text{chloride ion}}{Cl^-}$$

$$\underset{\text{acetic acid}}{CH_3\,COOH} \xrightarrow{H_2O} \underset{\text{acetate ion}}{CH_3\,COO^-} + \underset{\text{hydrogen ion}}{H^+}$$

$$\underset{\text{ammonium hydroxide}}{NH_4\,OH} \xrightarrow{H_2O} \underset{\text{ammonium ion}}{NH_4^+} + \underset{\text{hydroxyl ion}}{OH^-}$$

Some *strong electrolytes* always exist in ionic form even in the solid state, *e.g.,* sodium chloride (Na^+Cl^-), sodium bicarbonate ($Na^+HCO_3^-$). Other electrolytes dissociate completely into ions when in solution, *e.g.,* hydrogen chloride in water ($H^+ + Cl^-$). Besides, there are some *weak electrolytes,* which are only partially dissociated into ions, and their solutions contain undissociated molecules as well as the corresponding ions.

When ionic dissociation produces equal amounts of positive and negative electric charges, solutions containing dissociated ionic compounds remain electrically neutral. If placed in an electrical field, positively charged ions move towards the negative pole (Cathode). They are therefore called *cations.* On the other hand, negatively charged ions move towards the positive pole (anode); they are therefore called *anions.* Some physiologically important cations and anions are as follows:

Cations: hydrogen ion (H^+), ammonium ion (NH^+_4), metallic ions like Na^+, K^+, Ca^{2+}, Mg^{2+}, etc.

Anions: Chloride ion (Cl^-), hydroxyl ion (OH^-), bicarbonate ion (HCO_3^-), sulphate ion (SO_4^{2-}), dihydrogen phosphate ion ($H_2PO_4^-$), hydrogen phosphate ion (HPO_4^{2-}), etc.

It should be noted that many of the organic solutes in the body fluids like glucose and urea, are *non-electrolytes* and exist in solution only as unionised molecules. Water is also an excellent solvent for many non-electrolytes.

Role of Dissociation Constant

The acetic acid dissociates into ions according to this equation:

$$CH_3COOH \rightleftharpoons CH_3COO^- + H^+$$

The equilibrium constant for this reaction is:

$$K = \frac{[CH_3COO^-][H^+]}{[CH_3COOH]} = 1.78 \times 10^{-5} = 10^{-4.75}$$

In this case *K* describes the electrolytic dissociation of acetic acid and is called the *dissociation constant.* The higher the dissociation constant, the stronger the ionisation of the compound in question. Weak acids are those with small dissociation constants. As the exponential form is inconvenient, it is useful to express the dissociation of the weak acid in terms of the negative logarithm of the equilibrium constant.

$$\text{pK (acetic acid)} = -\log K = \log \frac{1}{K} = 4.75$$

All weak electrolytes have one or more pK values, which are written as pK_1, pK_2, etc. as shown below:

pK values of some electolytes

Acid	*Conjugate base*	*pK*
CH_3COOH	CH_3COO^-	4.75
H_3PO_4	$H_2PO_4^-$	2.0 (pK_1)
$H_2PO_4^-$	HPO_4^{2-}	6.7 (pK_2)
HPO_4^{2-}	PO_4^{3-}	11.7 (pK_3)
NH_4^+	NH_3	9.3
H_2CO_3	HCO_3^-	6.4 (pK_1)
HCO_3^-	CO_3^{2-}	10.2 (pK_2)

It should be noted that there is a special relationship between the *degree of dissociation* α (the value showing what that of the dissolved moles of the electrolyte has fallen into ions) and the dissociation constant. This relationship is described by the dilution law formulated by Ostwald:

$$\frac{\alpha^2}{1-\alpha}c = K$$

Where α is the degree of dilution; c is the molar concentration of the electrolyte dissociating into two ions; and K is the dissociation constant. Once two quantities in this equation are known, the third can be determined for the given conditions.

Such calculations give constant values of K only for weak electrolytes. For strong electrolytes K is not constant and increases with the concentration of the solution.

Dissociation degree and dissociation constant of electrolytes at 18°C

Electrolyte	*Concentration*	α, %	*K*
CH_3COOH	0.1N	1.32	1.76×10^{-5}
CH_3COOH	0.01N	4.10	1.75×10^{-5}
KC1	0.1N	86.20	0.536
KC1	0.01N	94.20	0.152

Since the dissociation constant of weak electrolytes is a constant value, the concentration of various ions in a solution can be changed.

Acids and their Characteristics

An acid is defined as any compound that can liberate a hydrogen ion.

$$\underset{\text{acetic acid}}{CH_3COOH} \rightarrow \underset{\text{acetate ion}}{CH_3COO^-} + \underset{\text{hydrogen ion}}{H^+}$$

This equation shows that *acetic acid* dissociates in such a way that hydrogen ions are formed. This is what actually makes acetic acid an acid. Some acids in aqueous solutions are fully ionised at all times. They are called *strong acids, e.g.,* hydrochloric acid.

$$HCl \rightleftharpoons H^+ + Cl^-$$

On the other hand, *weak acids* are only partially ionised, only a small proportion of their replaceable hydrogen being in the ionic form. For example,

$$\underset{\text{carbonic acid}}{H_2CO_3} \rightleftharpoons H^+ + HCO_3^-$$

$$\underset{\text{acetic acid}}{CH_3COOH} \rightleftharpoons H^+ + CH_3COO^-$$

In all the above cases H^+ ions are produced. But water breaks the bonds of different ion-producing compounds to different degrees. In the case of HCl, virtually all bonds of H are broken ; in the case of H_2CO_3 and CH_3COOH, only some are broken. Consequently, a solution of acetic acid will contain partly the whole compound CH_3COOH and partly the free acetate and hydrogen ions.

Bases and their Characteristics

A base is defined as any compound which can accept hydrogen ions or which is characterised by the ability to combine with the H^+ ions furnished by an acid. Some *strong bases* (alkalies), like sodium hydroxide, are completely ionised in solution at all times.

$$NaOH \longrightarrow Na^+ + OH^-$$

Thus, any compound which dissociates to yield *hydroxyl ions* (OH^-) is a base or an *alkali.* On the other hand, *weak bases* owe their basicity to their power to capture H^+ as shown below:

$$NH_3 + H^+ \rightleftharpoons NH_4^+$$

$$HPO_4^{2-} + H^+ \rightleftharpoons H_2PO_4^-$$

A solution of ammonia (ammonium hydroxide) is alkaline.

$$NH_3 + H^+ + OH^- \rightleftharpoons NH_4^+ + OH^-$$

It should be noted that acids and bases can be either neutral molecules (*e.g.*, HC1 and NaOH) or charged ions (for example, NH_4^+ - and HPO_4^{2-}). Most of the acids and bases of biological interest are generally weak.

Product of Acid and Base

A salt is defined as a compound that results from the chemical interaction of an acid and a base. Sodium chloride is a salt because it is formed by the interaction of hydrochloric acid and sodium hydroxide:

$$\underset{\text{acid}}{HC1} + \underset{\text{base}}{NaOH} \longrightarrow \underset{\text{salt}}{NaCl} + H_2O$$

Salts formed from a strong acid and a strong base are fully ionised even in the solid state, for example sodium chloride ($Na^+ Cl^-$). Salts from either a strong acid with a weak base or a weak acid with a strong base are also fully ionised when dissociated in water, but this does not mean that their solutions will be neutral. In the case of the salt of a weak acid with a strong base, the anion will react partially with H^+ from the water to give the undissociated weak acid, leaving an excess of OH^-. The solution will be alkaline with a pH above 7. For example, sodium bicarbonate in water.

$$Na^+ + HCO_3^- + H^+ + OH^- + H_2O \rightleftharpoons Na^+ + H_2CO_3 + OH^- + H_2O$$

Similarly the salt of a weak base with a strong acid gives an acid solution because of the reversible loss of H^+ from the cation. For example, ammonium chloride in water:

$$Cl^- + NH_4^+ + OH^- + H^+ + H_2O \rightleftharpoons Cl^- + NH_3 + H_2O + H^+ + H_2O$$

All salts of both strong and weak acids dissociate completely in water to free cations and anions.

pH (Hydrogen ion concentration): pH is defined as the negative logarithm of the hydrogen ion concentration in an aqueous solution (at a given temperature).

$$pH = -\log [H^+] \text{ moles/l}$$

$$= \log \frac{1}{[H^+]}$$

where the logarithms are taken to the base ten.

The following points describe the important characteristics of the pH scale:

1. The pH scale is inversely related to the hydrogen ion concentration $[H^+]$; that is, the lower the pH the higher the hydrogen ion concentration.
2. The pH scale is *exponential* rather than linear. A change of lpH unit corresponds to a 10-fold change of H^+ ion concentration of the solution. Thus, a change of pH from 7 to 6 represents a 10-fold increase in concentration of H^+ ions. Similarly a change from 7 to 8 represents a 10-fold increase in concentration of OH^- ions. In other words, pH 6 is 10 times more acidic than pH 7 and pH 5 is 100 times more acidic than pH 7 and so on, or pH 8 is 10 times more alkaline than pH 7 and pH 9 is 100 times more basic than pH 7.
3. For solutions of strong acids which dissociate

completely (such as HCl), the pH is approximately equal to the logarithm of the reciprocal of the acid concentration. Thus, the pH of a 0.001 M HCl solution is close to 3.

It follows that if the pH of a solution is 4, its hydrogen ion (H^+) concentration is 10^{-4} and its OH^- ion concentration is 10^{-10} at a given temperature. The pH scale is normally taken as extending from 0 to 14. A neutral solution has a pH value of 7, an acid solution has a pH value less than 7, and a basic solution has a pH value greater than 7 as indicated below:

		neutral	
pH	*acidic* →	↓ 7 ↑	*alkaline* →
	1 2 3 4 5 6		8 9 10 11 12 13 14
	Very acid (most H^+ released)	*Neutral*	Very basic (most OH^- released)

The reaction of neutral water and of any neutral solution is expressed as pH7

$$H^+ \times OH^- = K_w = 10^{-7} \times 10^{-7} = 10^{-14}$$
$$= K_w = 10^{-14}$$

(K_w means ionic product)

The concentration of H^+ and OH^- in neutral water is always 10^{-14}. Measurement shows that in a litre of pure water 0.0000001 g of H^+ ions is present. The pH of pure water therefore is:

pH (water) = log 1/0.0000001) = 7

Relationships of hydrogen ion concentrations are expressed in various ways. Living matter, containing a mixture of variously dissociated acids, bases, and salts, usually has a pH very near to neutrality. For example, the pH of human blood is generally 7.4. All enzymatic reactions are carried at an optimum pH.

Buffers

A buffer solution is one which can accommodate the

addition to it of moderate amounts of acid or base without marked changes in its pH value. The capacity of a solution to resist alteration in its pH value is known as *buffer capacity.*

The general form of the buffering reaction is:

$$\text{Buffer} + H^+ \rightleftharpoons \text{H Buffer}$$

In this example, a free H^+ combines with the buffer to form a weak acid (H Buffer) that can either remain as an undissociated molecule or dissociate back to Buffer and H^+. When the hydrogen ion concentration increases, the reaction is forced to the right and more hydrogen ions bind to the buffer, as long as available buffer is present. Conversely, when the hydrogen ion concentration decreases the reaction shifts towards the left and hydrogen ions are released from the buffer. Thus, changes in hydrogen ion concentration are minimised.

Relationships of H^+ ion Concentrations Expressed in Various Ways

Reaction	*Fraction of normality*	*H^+ions/l (grams)*	*Log of (H^+ions)*	*Expressed as pH*
Acid	N/1	1.0	-0	0.0
Acid	N/10	0.1	-1	1.0
Acid	N/100	0.01	-2	2.0
Acid	N/1000	0.001	-3	3.0
Acid	N/10,000	0.0001	-4	4.0
Acid	N/100,000	0.00001	-5	5.0
Acid	N/1000,000	0.000001	-6	6.0
Neutral	Pure Water	0.0000001	-7	7.0
Alkaline	N/1000,000	0.00000001	-8	8.0
Alkaline	N/100,000	0.000000001	-9	9.0
Alkaline	N/10,000	0.0000000001	-10	10.0
Alkaline	N/1000	0.00000000001	-11	11.0
Alkaline	N/100	0.000000000001	-12	12.0
Alkaline	N/10	0.0000000000001	-13	13.0
Alkaline	N/1	0.00000000000001	-14	14.0

Buffer System

A buffer system has the property of resisting pH changes despite additions of acid or base. An adequate buffer system can be obtained from a weak acid mixed with the salt from that acid and a strong base. For example,

$$\underset{\text{weak acid}}{HA} + \underset{\text{salt}}{B^+ A^-}$$

where B^+

= the cation of a strong base, *e.g.*, Na^+

A^- = anion from weak acid

Relationships of H^+ and OH^- ions and pH

H^+ ions	*OH^- ions*	*pH*	*Reaction*
10^{-0}	10^{-14}	0	Acid
10^{-1}	10^{-13}	1	"
10^{-2}	10^{-12}	2	"
10^{-3}	10^{-11}	3	"
10^{-4}	10^{-10}	4	"
10^{-5}	10^{-9}	5	"
10^{-6}	10^{-8}	6	"
10^{-7}	10^{-7}	7	Neutral
10^{-8}	10^{-6}	8	Alkaline
10^{-9}	10^{-5}	9	"
10^{-10}	10^{-4}	10	"
10^{-11}	10^{-3}	11	"
10^{-12}	10^{-2}	12	"
10^{-13}	10^{-1}	13	"
10^{-14}	10^{-0}	14	"

This corresponds to a partially neutralised solution of weak acid ($HA \rightarrow H^+ + A^-$) with some reasonable proportion of the acid converted to the anion. Its pH will be higher than the pH of the weak acid alone. The pH of a buffer solution is determined not on the absolute concentrations of buffer constituents but on their ratio, *i.e.*, on the ratio of the amount of salt to weak acid as stated in the *Henderson-Hasselbalch equation:*

$$pH = pK_a + \log\frac{[salt]}{[acid]}$$

where K_a is the dissociation constant of the acid and pK_a is the pH at which the acid is half neutralised and equal amounts of acid and ion (salt) are present. This equation is applicable to all the buffer systems. The buffer capacity, measured as the ability of the solution to minimise changes in pH due to addition of base, is strongest near the midpoint of a titration, when $[A^-] = [HA]$ and $pH = pK_a$. The most important examples of biological buffer systems are as follows:

Bicarbonate Buffer: The major buffer of extracellular fluid is the HCO_3^-/H_2CO_3 system. This results from a number of factors. (0 There is considerably more HCO_3^- present in extracellular fluid than any other buffer component, (ii) There is an unlimited supply of CO_2, (iii) Physiological mechanisms maintain normal extracellular pH by controlling either the HCO_3^-or the CO_2 concentration of extracellular fluid, and (iv) this buffer system operates in conjunction with Hb.

The Henderson-Hasselbalch equation for this buffer system may be written as:

$$pH = pK + \log\frac{[HCO^-{}_3]}{[H_2CO_3]}$$

In this case pK H_2CO_3 = 6.1, the ratio $\frac{[HCO^-{}_3]}{[H_2CO_3]} = \frac{20}{1}$ for

$$pH = pK + \log\frac{[HCO^-{}_3]}{[H_2CO_3]}$$

$$7.4 = 6.1 + 1.3$$

$$1.3 = \log 20$$

Moreover in true plasma total CO_2 content is equal to $[H_2CO_3]$ + $[HCO_3^-]$. Hence, if the total CO_2 content of true plasma is known and the pCO_2 of alveolar air, *i.e.*, arterial pCO_2 is determined, the pH of the true plasma can be calculated. This method of determining blood pH (true plasma pH) was used before the advent of commercial pH metres. The Henderson-Hasselbalch equation is important, for it enables us to appreciate that the blood pH is dependent on the *ratio* of the concentration of free acid to the concentration of the buffer anion (bicarbonate). Therefore, acidaemia or alkalaemia could be caused by either a change in the non-volatile component (HCO_3^- concentration) or in the volatile component (H_2CO_3 concentration and hence pCO_2).

Since the $[H_2CO_3]$ is fixed only by the alveolar CO_2 tension, if the gas tension is equivalent to that normally present in the blood, the HCO_3–H_2CO_3 system is more useful at pH 7.4, with a ratio of 20, than it would be at its pK 6.1 where the HCO_3^- would be exhausted by addition of 1.25 meq per litre of acid, and addition of this amount of alkali would result in a rise of 0.3 pH unit.

The buffer efficiency of HCO_3^-/H_2CO_3 is further enhanced by the presence of erythrocytes. As CO_2 diffuses into cells, the H_2CO_3 reacts with Hb forming HCO_3-, which then enters the plasma in exchange for Cl^-. This is not contingent upon deoxygenation of Hb, but is achieved more readily and with even less pH change when deoxygenation occurs simultaneously. Conversely, lowering the CO_2 tension results in a reversal of this process, with consequent diminution of plasma $[HCO_3$-]. Although the situation does not arise under physiological conditions, only in the presence of red cells does the total CO_2 content of plasma fall to zero at a pCO_2 of zero mm. This is possible because there is sufficient Hb in whole blood to permit the following series of reactions.

$$H^+ + HbO_2 \longrightarrow O_2 + HHb^+$$

$$HHb^+ + HCO_3^- \rightleftharpoons Hb + H_2CO_3 \rightleftharpoons H_2O + CO_2$$

The normal value of 20: 1 of the ratio of HCO_3-/H_2CO_3 is also maintained through respiratory regulation of the pH of extracellular fluid. In addition to the automatic self adjustments made possible by intracorpuscular Hb (as described above), the body possesses two further safeguards (the respiratory apparatus and the kidneys) which by their control of plasma [H_2CO_3] and [HCO_3^-] respectively serve in auxiliary fashion to maintain constant pH of blood plasma. Unlike the [HCO_3^-] the [H_2CO_3] is determined solely by the partial pressure of CO_2 in the alveolar air in equilibrium with extracellular fluid.

This, in turn, is dependent upon the rate at which CO_2 leaving pulmonary blood is diluted with atmospheric air and hence upon the rate and depth of respiration. The latter are regulated by the nervous system at the respiratory centre which is sensitive to pH and pCO_2 of the extracellular fluid. When the pH falls below the normal because of diminished [HCO_3^-], respiration is stimulated, lowering alveolar pCO_2 and hence extracellular [H_2CO_3]. This tends to return the HCO_3–H_2CO_3 ratio to its normal value of 20: 1 and thus to restore pH towards 7.4. With a high plasma pH, the respiratory rate falls, alveolar pCO_2 and hence plasma [H_2CO_3] rise, and the pH moves towards 7.4.

But perfect compensation is not attained since the increased plasma [H_2CO_3] opposes the effect of elevated pH on the respiratory centre. If the respiratory rate falls sufficiently, the diminished pO_2, sensed by the carotid body chemoreceptors, could be a stimulus for increased respiratory activity.

As mentioned earlier the pH is dependent not on absolute concentrations but solely on the HCO_3^-/H_2CO_3 ratio. The buffer systems of plasma can withstand the addition of 16 meq of acid or 29 meq of alkali per litre and still maintain the pH within the range compatible with life. With pulmonary compensation the normal pH range can be maintained despite addition of 23 meq of acid or 80 meq of alkali per litre of plasma.

Phosphate Buffer: Although the contribution of phosphate buffer ($HPO_4^{2-}/H_2PO_4^-$) to the buffering power of the plasma is negligible because their plasma concentration is very low (3 mg/100ml), this buffer system is present (as is HCO_3^-/H_2CO_3) in the tubular fluid of the kidney. The Henderson-Hasselbalch equation for this buffer system may be written as :

$$pH = pK + \log \frac{[salt]}{[acid]}$$

$$pH = pK_{phos} + \log \frac{[HPO_4^{2-}]}{[H_2PO_4^-]}$$

$$7.4 = 6.8 + 0.6$$

$$0.6 = \log 4$$

Thus, phosphate equilibrium in plasma is such that 4 parts of the so-called *basic phosphate* (HPO_4^{2-}) exist to 1 part of *acidic phosphate* ($H_2PO_4^-$) at a pH of 7.4, for pK_{phos} is 6.8.

At a urinary pH of 5.8, ten parts of phosphate are in the 'acidic' form and one part in the 'basic' form (5.8 = 6.8 +1⁻), for each phosphate ion excreted as $H_2PO_4^-$ one Na^+ ion is saved and one H^+ ion is excreted. The phosphates are by far the most important buffers in the urine, although in acid urine the formation of NH_4 from NH_3 and H^+ helps to keep the $[H^+]$ lower than it would otherwise be.

Thus, phosphate buffer system contributes little to the total buffering capacity of blood because the blood phosphate concentration is very low compared with the amount of protein present. But the phosphate buffer pair is a major outlet for H^+ via the urine which has a relatively high phosphate content. Hence the pH of the urine is provided by the excretion of H^+ into the tubular fluid.

Oxyhaemoglobin Buffer: The buffer pair formed by HbO_2^- /$H.HbO_2$, *i.e.*, oxyhaemoglobin anion/oxyhaemoglobin (acid) is also important. It is well known that uptake of CO_2 from the tissue depends upon the following reactions:

$$CO_2 + H_2O \rightleftharpoons H_2CO_3 \rightleftharpoons H^+ + HCO_3^-$$

Thus, the uptake of CO_2 from the tissues is dependent upon the supply of hydrogen acceptors so that the reaction can proceed towards the right. The most important hydrogen acceptor in blood is the haemoglobin ion. Most of the power of haemoglobin in mopping up hydrogen ions in the range pH 7.0 to 7.70 comes from the dissociation of its imidazole groups [haemoglobin contains 38 molecules of histidine (β-imidazole α-amino-propionic acid) per molecule]. The dissociation tendency of the imidazole group like that of any weak acid is largely dependent upon the pH of the solution. As the pH rises, the dissociation is reduced and as a result the H^+ ions added are 'mopped up'.

Oxyhaemoglobin dissociates more completely than does reduced haemoglobin. As a result, reduced haemoglobin produces less H^+ at a given pH than does oxyhaemoglobin. Thus, when oxyhaemoglobin solution is reduced the pH of the solution falls, *i.e., reduced haemoglobin* is a better H^+ acceptor than oxyhaemoglobin. This is of great importance in blood physiology because the entrance of CO_2 from the tissues to the capillary blood is accompanied by the simultaneous reduction of oxyhaemoglobin. As the uptake of CO_2 depends upon H^+ acceptors, this increase in H^+ acceptance by the reduced form facilitates the uptake and buffering of CO_2. For each mMol of oxyhaemoglobin reduced, about 0.7 mMol of H^+ can be taken up ; as a result 0.7 mMol CO_2 can enter the blood without any change in pH.

As arterial blood enters the tissues, CO_2 diffuses into the erythrocytes, thus potentially lowering the pH and the affinity of Hb for O_2 as shown in the following reactions.

$$CO_2 + H_2O \rightleftharpoons H_2CO_3 \rightleftharpoons H^+ + HCO_3^-$$

$$H^+ + HbO_2 \rightarrow O_2 + HHb^+$$

On the other hand, in the lungs the loss of CO_2, which

would potentially raise the pH, increases the affinity of the Hb for O_2, thus permitting the saturation of Hb with O_2 at a lower pO_2.

$$HHb^+ + O_2 \rightleftharpoons HbO_2 + H^+$$

It should be noted that oxygenation of Hb results in a shift in the apparent pK_a of some acidic groups on the peptide chains from 7.71 to 6.17; thus HbO_2 acts as a stronger acid than Hb. Approximately 0.7 mol of H^+ is released as 1 equivalent of O_2 is bound.

Importance of Buffer Systems

Buffer mixtures are very important in living organisms and the mineral world. An example of a natural buffer is the blood of mammals. It always contains free carbonic acid and sodium carbonate. Therefore the pH of blood is always maintained at 7.4.

The buffering action of soils is very important in agriculture, because plants absorb artificial fertilizers from the soil to change the pH in solutions that they extract from the soil in an unfavourable direction. An imbalance in the buffering action of soil is detrimental to useful micro-organisms living in it.

Buffer solutions are very important in the treatment of domestic sewage, because the micro-organisms which mineralise their organic matter thrive better in a neutral medium. A shift towards acidity or alkalinity inhibits the vital processes in the microbes, thus adversely affecting the working of sewage treatment plants.

Buffers play an important role in the chemical treatment of water to separate it from suspended matter of coagulation. The higher the buffer capacity of the treated water, the more efficient its purification with a hydrolysing coagulant. The buffer capacity of natural water accounts for its neutralising power.

Buffers are widely used in volumetric analysis. For example, an ammonium buffer (HH_4OH + HH_4Cl) is used to determine Ca and Mg ions in water (trilonometric method). Buffer mixtures are used to determine the pH of solutions colorimetrically and potentiometrically.

In cellular physiology, buffers are an important way for the cell to maintain constant or smoothly changing conditions. Since cellular metabolism is constantly producing and consuming protons, the pH of the cell in the absence of buffers would change rapidly between low and high levels.

7

Consumption of Energy

The energy is used for all the processes which keep the organism alive. Some of these processes occur continually, such as the metabolism of foods, the synthesis of large, biologically important molecules, e.g. proteins and DNA, and the transport of molecules and ions throughout the organism. Other processes occur only at certain times, such as muscle contraction and other cellular movements.

Animals obtain their energy by oxidation of foods, plants do so by trapping the sunlight using chlorophyll. However, before the energy can be used, it is first transformed into a form which the organism can handle easily. This special carrier of energy is the molecule adenosine triphosphate, or ATP.

Cellular Respiration

The process occurs in two phases:

— glycolysis, the breakdown of glucose to pyruvic acid
— the complete oxidation of pyruvic acid to carbon dioxide and water

In eukaryotes, glycolysis occurs in the cytosol. The remaining processes take place in mitochondria.

Mitochondria: Mitochondria are membrane-enclosed organelles distributed through the cytosol of most eukaryotic cells. Their main function is the conversion of the potential energy of food molecules into ATP. Mitochondria have:

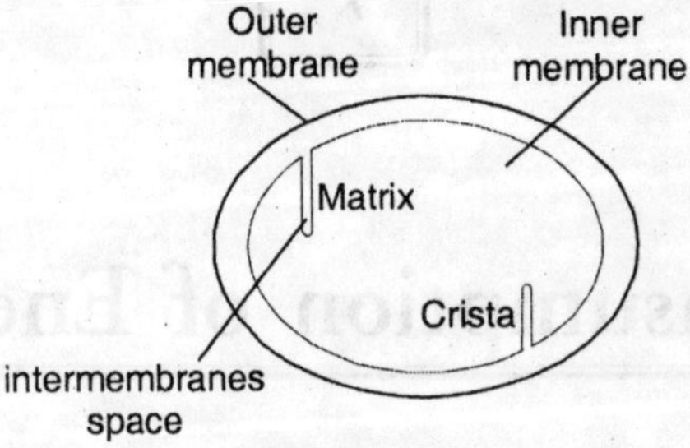

Mitochondria

— an outer membrane that encloses the entire structure
— an inner membrane that encloses a fluid-filled matrix
— between the two is the intermembrane space
— the inner membrane is elaborately folded with shelflike cristae projecting into the matrix.
— a small number (some 5-10) circular molecules of DNA

Outer Membrane: The outer membrane contains many complexes of integral membrane proteins that form channels through which a variety of molecules and ions move in and out of the mitochondrion.

Inner Membrane: The inner membrane contains 5 complexes of integral membrane proteins:

— NADH dehydrogenase (Complex I)
— succinate dehydrogenase (Complex II)
— cytochrome c reductase (Complex III; also known as the cytochrome b-C_1 complex)
— cytochrome c oxidase (Complex IV)
— ATP synthase (Complex V)

Matrix

— oxidised by NAD^+ producing NADH+ H^+

— decarboxylated producing a molecule of
— carbon dioxide (CO_2) and
— a 2-carbon fragment of acetate bound to coenzyme A forming acetyl-CoA

Citric Acid Cycle

— This 2-carbon fragment is donated to a molecule of oxaloacetic acid.
— The resulting molecule of citric acid (which gives its name to the process) undergoes the series of enzymatic steps.
— The final step regenerates a molecule of oxaloacetic acid and the cycle is ready to turn again.
— Each of the 3 carbon atoms present in the pyruvate that entered the mitochondrion leaves as a molecule of carbon dioxide (CO_2).
— At 4 steps, a pair of electrons ($2e^-$) is removed and transferred to NAD^+ reducing it to $NADH + H^+$.
— At one step, a pair of electrons is removed from succinic acid and reduces FAD to $FADH_2$.

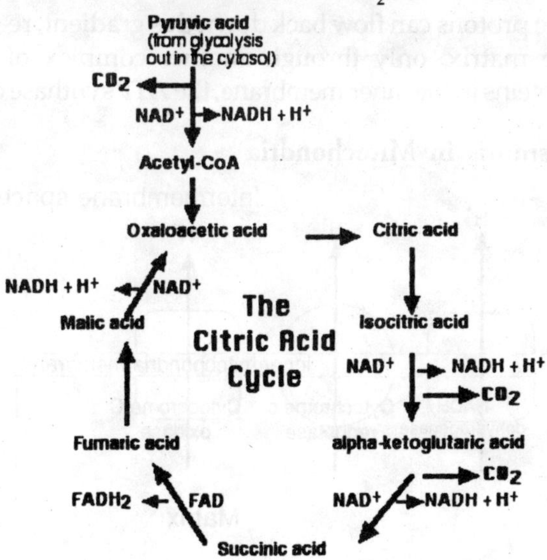

Respiratory Chain

Respiratory Chain: The respirintegral membrane proteins chain consists of 3 complexes of integral membrane proteins

— the NADH dehydrogenase complex (I)
— the cytochrome c reductase complex (III)
— the cytochrome c oxidase complex (IV)

and two freely-diffusible molecules

— ubiquinone
— cytochrome c

that shuttle electrons from one complex to the next.

— the stepwise transfer of electrons from NADH (and $FADH_2$) to oxygen molecules to form (with the aid of protons) water molecules (H_2O);
— harnessing the energy released by this transfer to the pumping of protons (H^+) from the matrix to the intermembrane space.
— Approximately 20 protons are pumped into the intermembrane space as the 4 electrons needed to reduce oxygen to water pass through the respiratory chain.
— The gradient of protons formed across the inner membrane by this process of active transport forms a miniature battery.
— The protons can flow back down this gradient, re-entering the matrix, only through another complex of integral proteins in the inner membrane, the ATP synthase complex.

Chemiosmosis in Mitochondria

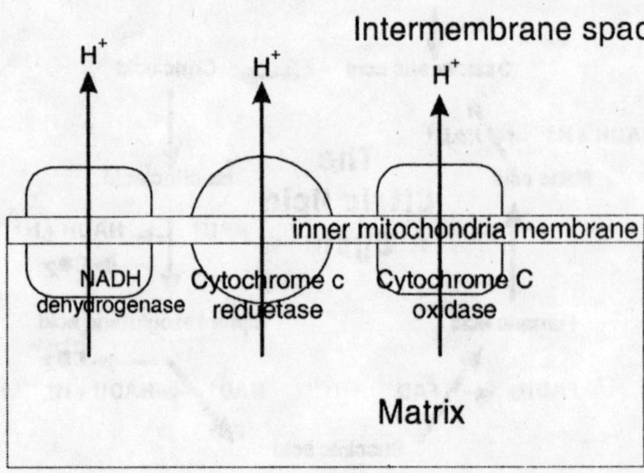

ATPs

— 10 by NAD^+ (so 10x3=30); and
— 2 by $FADH_2$ (so 2x2=4), this could generate 34 ATPs.
— The energy stored in the proton gradient is also used for the active transport of several molecules and ions through the inner mitochondrial membrane into the matrix.
— NADH is also used as reducing agent for many cellular reactions.

Three Exceptions: A stoichiometric production of ATP does occur at:

— one step in the citric acid cycle yielding 2 ATPs for each glucose molecule. This step is the conversion of alpha-ketoglutaric acid to succinic acid.
— at two steps in glycolysis yielding 2 ATPs for each glucose molecule.

Mitochondrial DNA (mtDNA)

— 2 different molecules of ribosomal RNA (rRNA)
— 22 different molecules of transfer RNA (tRNA) (at least one for each amino acid)
— 13 polypeptides

Mutations in mtDNA cause human diseases. A number of human diseases are caused by mutations in genes in our mitochondria:

— cytochrome b
— 12S rRNA
— ATP synthase
— subunits of NADH dehydrogenase
— several tRNA genes

A number of humans who suffer from easily-fatigued muscles turn out to have a mutations in their cytochrome b gene. Curiously, only the mitochondria in their muscles have the mutation; the mtDNA of their other tissues is normal. Presumably, very early in their embryonic development, a mutation occurred in the cytochrome b gene in a cell destined to produce their muscles.

Adenosine Triphosphate (ATP): ATP is regarded as a universal source of energy occurring in all cell types. It is produced mainly during the oxidation of energy-rich (reduced) compounds processed in the respiratory chain and in photosynthesis. ATP is needed

— as a source of energy for biochemical syntheses
— for transport processes (active transport) and
— for mechanical work like movements (ciliar movements, plasma currents, etc.)

ATP is the nucleotide known in biochemistry as the "molecular currency" of intracellular energy transfer; that is, ATP is able to store and transport chemical energy within cells. ATP also plays an important role in the synthesis of nucleic acids. ATP molecules are also used to store the usable energy that plants convert in cellular respiration.

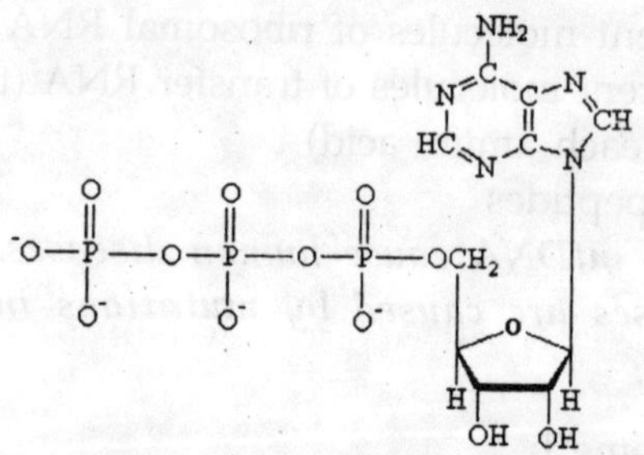

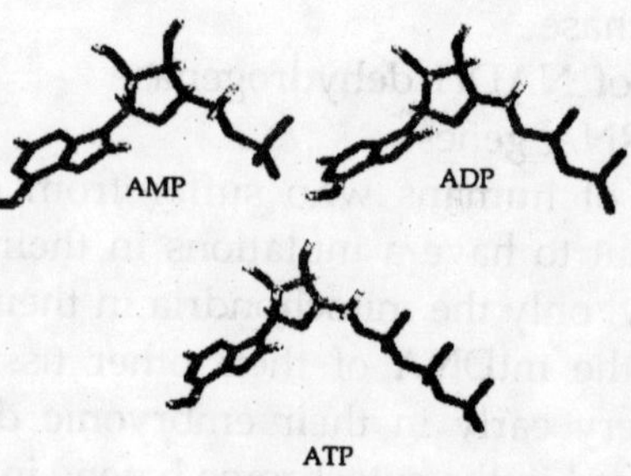

The ATP molecule is composed of three components. At

the centre is a sugar molecule, ribose (the same sugar that forms the basis of DNA). Attached to one side of this is a base (a group consisting of linked rings of carbon and nitrogen atoms); in this case the base is adenine. The other side of the sugar is attached to a string of phosphate groups. These phosphates are the key to the activity of ATP.

ATP works by losing the endmost phosphate group when instructed to do so by an enzyme. This reaction releases a lot of energy, which the organism can then use to build proteins, contact muscles, etc. The reaction product is adenosine diphosphate (ADP), and the phosphate group either ends up as orthophosphate (HPO_4) or attached to another molecule. Even more energy can be extracted by removing a second phosphate group to produce adenosine monophosphate (AMP).

$$ATP + H_2O \rightarrow ADP + HPO_4$$

When the organism is resting and energy is not immediately needed, the reverse reaction takes place and the phosphate group is re-attached to the molecule using energy obtained from food or sunlight. Thus the ATP molecule acts as a chemical 'battery', storing energy when it is not needed, but able to release it instantly when the organism requires it.

Phosphorus Cycle: The fact that ATP is Nature's 'universal energy store' explains why phosphates are a vital ingredient in the diets of all living things. Modern fertilizers often contain phosphorus compounds that have been extracted from animal bones. These compounds are used by plants to make ATP. We then eat the plants, metabolise their phosphorus, and produce our own ATP. When we die, our phosphorus goes back into the ecosystem to begin the cycle again.

Chemical Properties: Chemically, ATP consists of adenosine and three phosphate groups (triphosphate). It has the empirical formula $C_{10}H_{16}N_5O_{13}P_3$, and the chemical formula $C_{10}H_8N_4O_2NH_2(OH)_2(PO_3H)_3H$. The phosphoryl groups starting with that on AMP are referred to as the alpha (a), beta (B), and gamma (y) phosphates. The biochemical names for ATP are 9-p-D-ribofuranosyladenine-5'-triphosphate and, equivalently, 9-B-D-ribofuranosyl-6-amino-purine-5'-triphosphate.

ATP can be produced by various cellular processes, most typically in mitochondria by oxidative phosphorylation under the catalytic influence of ATP synthase or in the case of plants in chloroplasts by photosynthesis. The main fuels for ATP synthesis are glucose and fatty acids. Initially glucose is broken down into pyruvate in the cytosol. Two molecules of ATP are generated for each molecule of glucose. The terminal stages of ATP synthesis are carried out in the mitochondrion and can generate up to 36 ATP.

The total quantity of ATP in the human body is about 0.1 mole. The energy used by human cells requires the hydrolysis of 200 to 300 moles of ATP daily. This means that each ATP molecule is recycled 2000 to 3000 times during a single day. ATP cannot be stored, hence its consumption must closely follow its synthesis.

Energy is stored in the covalent bonds between phosphates, with the greatest amount of energy (approximately 7 kcal/mole) in the bond between the second and third phosphate groups. This covalent bond is known as a pyrophosphate bond. We can write the chemical reaction for the formation of ATP as:

— *in chemicalese:* ADP + Pi + energy $\rightarrow$ ATP

— *in English:* Adenosine diphosphate + inorganic Phosphate + energy produces

Adenosine Triphosphate

The chemical formula for the expenditure/release of ATP energy can be written as:

— in chemicalese: ATP $\rightarrow$ ADP + energy + Pi

— in English Adenosine Triphosphate produces Adenosine diphosphate + energy + inorganic Phosphate

An analogy between ATP and rechargeable batteries is appropriate. The batteries are used, giving up their potential energy until it has all been converted into kinetic energy and heat/unusable energy. Recharged batteries (into which energy has been put) can be used only after the input of additional energy.

Thus, ATP is the higher energy form (the recharged battery) while ADP is the lower energy form (the used battery). When the terminal (third) phosphate is cut loose, ATP becomes ADP (Adenosine diphosphate; di= two), and the stored energy is released for some biological process to utilise. The input of additional energy (plus a phosphate group) "recharges" ADP into ATP (as in my analogy the spent batteries are recharged by the input of additional energy).

Two processes convert ADP into ATP:

— substrate-level phosphorylation; and
— chemiosmosis.

Substrate-level phosphorylation occurs in the cytoplasm when an enzyme attaches a third phosphate to the ADP (both ADP and the phosphates are the substrates on which the enzyme acts).

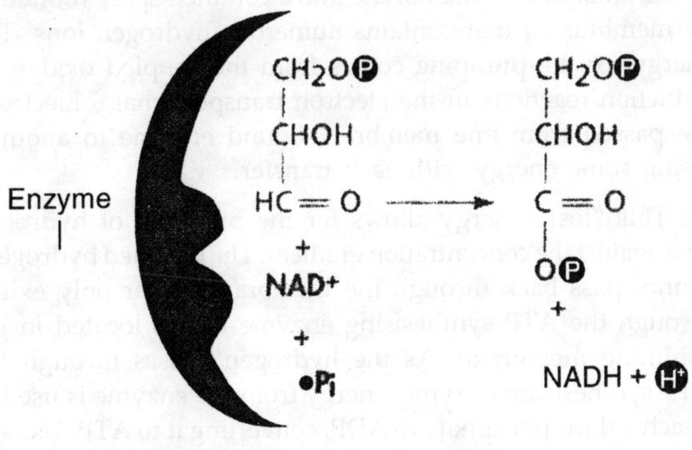

Enzymes and the formation of NADH and ATP.

Chemiosmosis involves more than the single enzyme of substrate-level phosphorylation. Enzymes in chemiosmotic synthesis are arranged in an electron transport chain that is embedded in a membrane. In eukaryotes this membrane is in either the chloroplast or mitochondrion. According to the chemiosmosis hypothesis proposed by Peter Mitchell in 1961, a special ATP-synthesising enzyme is also located in the membranes.

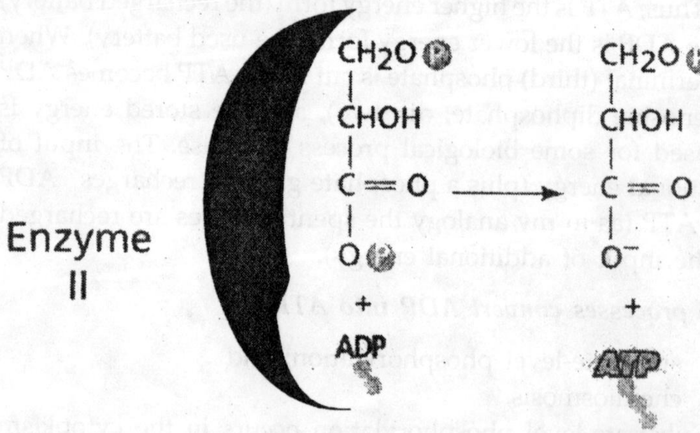

Typical representation of an electron transport chain.

During chemisomosis in eukaryotes, H^+ ions are pumped across an organelle membrane into a confined space (bounded by membranes) that contains numerous hydrogen ions. The energy for the pumping comes from the coupled oxidation-reduction reactions in the electron transport chain. Electrons are passed from one membrane-bound enzyme to another, losing some energy with each transfer.

This "lost" energy allows for the pumping of hydrogen ions against the concentration gradient. The confined hydrogens cannot pass back through the membrane. Their only exit is through the ATP synthesising enzyme that is located in the confining membrane. As the hydrogen passes through the ATP synthesising enzyme, energy from the enzyme is used to attach a third phosphate to ADP, converting it to ATP. Usually the terminal phosphate is not simply removed, but instead is attached to another molecule. This process is known as phosphorylation.

— W + ATP → W~P + ADP where W is any compound, for example:

— glucose + ATP → glucose~P + ADP

Glucose can be converted into Glucose-6-phosphate by the addition of the phosphate group from ATP.

ATP serves as the biological energy company, releasing energy for both anabolic and catabolic processes and being recharged by energy generated from other catabolic reactions.

Glycolysis

Cellular respiration is, in its broadest definition, the process in which the chemical bonds of energy-rich molecules such as glucose are converted into energy usable for life processes. Oxidation of organic material-in a bonfire, for example-is an exothermic reaction that releases a large amount of energy rather quickly. The overall equation for the oxidation of glucose is:

$C_6H_{12}O_6 + 6O_2 \rightarrow 6CO_2 + 6H$<= = Glycolysis = =

Glycolysis is a metabolic pathway that is found in all living organisms and does not require oxygen. The process converts one molecule of glucose into two molecules of pyruvate, and makes energy in the form of two molecules of ATP. Glycolysis takes place in the cytoplasm of the cell.

Glycolysis involves a series of biochemical reactions by which a molecule of glucose is oxidised to two molecules of pyruvic acid. In addition to the conversion of glucose to pyruvic acid, two molecules of oxidised NAD are reduced, two molecules of adenosine diphosphate (ADP) are phosphorylated to adenosine triphosphate (ATP) molecules, and two molecules of water and two protons (H^+) are produced.

Glycolysis is the initial process of many pathways of

carbohydrate catabolism, and serves two principal functions: generation of high-energy ATP molecules, and production of a variety of 6- or 3-carbon intermediate metabolites which may be removed at various steps in the process for other intracellular purposes. Glycolysis is one of the most universal metabolic processes known, and occurs (with minor variations) in many types of cells in nearly all types of organisms from bacteria to plants to animals and humans.

Although glycolysis produces less energy per glucose molecule than complete aerobic oxidation, it can occur at great speed and is anaerobic. The most common and well-known form of glycolysis is the Embden-Meyerhof pathway, initially elucidated by Gustav Embden and Otto Meyerhof. The term can be taken to include alternative pathways, such as the Entner-Doudoroff Pathway. However, glycolysis will be used here as a synonym for the Embden-Meyerhof pathway.

Steps of Glycolysis

Output: Glycolysis converts one molecule of glucose into two molecules of pyruvate, along with "reducing equivalents" in the form of the co-enzyme NADH. The overall reaction of glycolysis is:

$$\text{Glucose} + 2\ NAD^+ + 2\ ADP + 2\ P_i \rightarrow 2\ NADH + 2\ \text{pyruvate} + 2\ ATP + 2\ H_2O + 2\ H^+$$

So, for simple fermentations, the metabolism of 1 molecule of glucose has a net yield of 2 molecules of ATP. Cells performing respiration synthesise much more ATP but this is not considered part of glycolysis. Eukaryotic aerobic respiration produces an additional 34 molecules (approximately) of ATP for each glucose molecule oxidised. Unlike those molecules of ATP produced by aerobic respiration, those of glycolysis are produced by substrate-level phosphorylation.

Location: In eukaryotes glycolysis takes place within the cytosol of the cell. Glucose enters the cell through facilitated diffusion. In many animal tissues, including skeletal muscle, insulin stimulates this process. The ultimate fate of the pyruvate and NADH produced in glycolysis depends upon the organism and the conditions, most notably the presence or absence of oxygen or other external electron acceptors.

In fermentation, the pyruvate and NADH are anaerobically metabolised to yield any of a variety of products with an organic molecule acting as the final electron acceptor. For example, the bacteria involved in making yogurt simply reduce the pyruvate to lactic acid, whereas yeast produce ethanol and carbon dioxide.

In aerobic organisms, the pyruvate typically enters the citric acid cycle (also known as the TCA or Krebs cycle), and the NADH is ultimately oxidised by oxygen during oxidative phosphorylation. Although human metabolism is primarily aerobic, under anaerobic conditions, for example in over-worked muscles that are starved of oxygen, pyruvate is converted to lactate, as in many microorganisms.

Evolution: Glycolysis is highly conserved in evolution, being common to nearly all living organisms. This suggests great antiquity; it may have originated with the very first prokaryotes, 3.5 billion years ago or more.

Pathway: The first step in glycolysis is phosphorylation of glucose by hexokinase to form glucose 6-phosphate (G-6-P). In the liver an isozyme of hexokinase called glucokinase is used, which differs primarily in regulatory properties. This reaction consumes 1 ATP molecule, but the energy is well spent. The cell membrane is permeable to glucose for two reasons -glucose transporters move glucose across the membrane, and neutrally charged glucose is able to passively diffuse through. G-6-P is negatively charged and this is repelled by the plasma membrane, so this phosphorylation effectively traps glucose in the cell.

G-6-P is then rearranged into fructose 6-phosphate (F-6-P) by phosphoglucose isomerase. Fructose can also enter the glycolytic pathway via phosphorylation at this point. Phosphofructokinase-1 then consumes 1 ATP to form fructose 1,6-bisphosphate (F-l,6-bisP). The energy expenditure in this step is justified in 2 ways: the glycolytic process (up to this step) is now irreversible, and the energy supplied destablises the molecule, allowing the ring to be split by aldolase into dihydroxyacetone phosphate and glyceraldehyde 3-phosphate.

Triose phosphate isomerase rapidly interconverts these, with glyceraldehyde 3-phosphate proceeding further into glycolysis. Each molecule of glyceraldehyde 3-phosphate is then oxidised by a molecule of NAD^+ in the presence of glyceraldehyde 3-phosphate dehydrogenase, forming 1, 3-bisphosphoglycerate.

In the next step, phosphoglycerate kinase generates a molecule of ATP while forming 3-phosphoglycerate. At this step glycolysis has reached the break-even point: 2 molecules of ATP were consumed, and 2 new molecules have been synthesised. This step, one of the two substrate-level phosphorylation steps, requires ADP; thus, when the cell has plenty of ATP (and little ADP) this reaction does not occur.

Because ATP decays relatively quickly when it is not metabolised, this is an important regulatory point in the glycolytic pathway. Phosphoglyceromutase then forms 2-phosphoglycerate; enolase then forms phosphoenolpyruvate; and another substrate-level phosphorylation then forms a molecule of pyruvate and a molecule of ATP by means of the enzyme pyruvate kinase. This serves as an additional regulatory step.

After the formation of fructose 1,6 bisphosphate, many of the reactions are energetically unfavourable. The only reactions that are favourable are the 2 substrate-level phosphorylation steps that result in the formation of ATP. These two reactions pull the glycolytic pathway to completion.

All organisms have glycolysis occurring in their cytoplasm. At steps 1 and 3 ATP is converted into ADP, inputting energy into the reaction as well as attaching a phosphate to the glucose. At steps 6 and 9 ADP is converted into the higher energy ATP. At step 5 NAD^+ is converted into NADH + H^+.

The process works on glucose, a 6-C, until step 4 splits the 6-C into two 3-C compounds. Glyceraldehyde phosphate (GAP, also known as phosphoglyceraldehyde, PGAL) is the more readily used of the two. Dihydroxyacetone phosphate can be converted into GAP by the enzyme Isomerase. The end of the glycolysis process yields two pyruvic acid (3-C) molecules, and a net gain of 2 ATP and two NADH per glucose.

Under anaerobic conditions, the absence of oxygen, pyruvic acid can be routed by the organism into one of three pathways: lactic acid fermentation, alcohol fermentation, or cellular (anaerobic) respiration. Humans cannot ferment alcohol in their own bodies, we lack the genetic information to do so.

These biochemical pathways, with their myriad reactions catalysed by reaction-specific enzymes all under genetic control, are extremely complex. We will only skim the surface at this time and in this course. Alcohol fermentation is the formation of alcohol from sugar. Yeast, when under anaerobic conditions, convert glucose to pyruvic acid via the glycolysis pathways, then go one step farther, converting pyruvic acid into ethanol, a C-2 compound.

Many organisms will also ferment pyruvic acid into, other chemicals, such as lactic acid. Humans ferment lactic acid in muscles where oxygen becomes depleted, resulting in localised anaerobic conditions. This lactic acid causes the muscle stiffness couch-potatoes feel after beginning exercise programmes. The stiffness goes away after a few days since the cessation of strenuous activity allows aerobic conditions to return to the muscle, and the lactic acid can be converted into ATP via the normal aerobic respiration pathways.

Aerobic Respiration: When oxygen is present (aerobic conditions), most organisms will undergo two more steps, Kreb's Cycle, and Electron Transport, to produce their ATP.

In eukaryotes, these processes occur in the mitochondria, while in prokaryotes they occur in the cytoplasm.

Acetyl Co-A: Pyruvic acid is first altered in the transition reaction by removal of a carbon and two oxygens (which form carbon dioxide). When the carbon dioxide is removed, energy is given off, and NAD^+ is converted into the higher energy form NADH. Co-enzyme A attaches to the remaining 2-C (acetyl) unit, forming acetyl Co-A. This process is a prelude to the Kreb's Cycle.

Kreb's Cycle: The Acetyl Co-A (2-C) is attached to a 4-C chemical (oxaloacetic acid). The Co-A is released and returns to await another pyruvic acid. The 2-C and 4-C make another chemical known as Citric acid, a 6-C. Kreb's Cycle is also known as the Citric Acid Cycle. The process after Citric Acid is essentially removing carbon dioxide, getting out energy in the form of ATP, GTP, NADH and $FADH_2$, and lastly regenerating the cycle.

Between Isocitric Acid and a-Ketoglutaric Acid, carbon dioxide is given off and NAD^+ is converted into NADH. Between a-Ketoglutaric Acid and Succinic Acid the release of carbon dioxide and reduction of NAD+ into NADH happens again, resulting in a 4-C chemical, succinic acid.

GTP (Guanine Triphosphate, which transfers its energy to ATP) is also formed here (GTP is formed by attaching a phosphate to GDP). The remaining energy carrier-generating steps involve the shifting of atomic arrangements within the 4-C molecules. Between Succinic Acid and Fumaric Acid, the molecular shifting releases not enough energy to make ATP or NADH outright, but instead this energy is captured by a new energy carrier, Flavin adenine dinucleotide (FAD).

FAD is reduced by the addition of two H's to become $FADH_2$. $FADH_2$ is not as rich an energy carrier as NADH, yielding less ATP than the latter. The last step, between Malic Acid and Oxaloacetic Acid reforms OA to complete the cycle. Energy is given off and trapped by the reduction of NAD+ to

NADH. The carbon dioxide released by cells is generated by the Kreb's Cycle, as are the energy carriers (NADH and $FADH_2$) which play a role in the next step.

Electron Transport Phosphorylation: Whereas Kreb's Cycle occurs in the matrix of the mitochondrion, the Electron Transport System (ETS) chemicals are embedded in the membranes known as the cristae. Kreb's cycle completely oxidised the carbons in the pyruvic acids, producing a small amount of ATP, and reducing NAD and FAD into higher energy forms.

In the ETS those higher energy forms are cashed in, producing ATP. Cytochromes are molecules that pass the "hot potatoes" (electrons) along the ETS chain. Energy released by the "downhill" passage of electrons is captured as ATP by ADP molecules. The ADP is reduced by the gain of electrons. ATP formed in this way is made by the process of oxidative phosphorylation.

The mechanism for the oxidative phosphorylation process is the gradient of H+ ions discovered across the inner mitochondrial membrane. This mechanism is known as chemiosmotic coupling. This involves both chemical and transport processes. Drops in the potential energy of electrons moving down the ETS chain occur at three points. These points turn out to be where ADP + P are converted into ATP. Potential energy is captured by ADP and stored in the pyrophosphate bond. NADH enters the ETS chain at the beginning, yielding 3 ATP per NADH. $FADH_2$ enters at Co-Q, producing only 2 ATP per $FADH_2$.

Electron Transport and Oxidative Phosphorylation

The loss of substance that is generated incidentally by the respiration has the aim to develop mechanical forces by which the atoms or molecules of the other substances are set in those motions that result in growth and other functions of the living plant.

Respiration is the source that fuels all phenomenons of life while the assimilation occurring in chlorophyll- containing organs generates the substances that are subsequently set in motion to serve life. All reactions of glycolysis take place in the cell's cytosol. The citric acid cycle and the respiratory chain are located in another compartment: within mitochondria at the inner mitochondrial membrane.

It is here that ATP is generated under consumption of oxygen and this is why mitochondria are often said to be the cell's 'power plant'. The process is also called oxidative phosphorylation. It resembles Photophosphorylation that is part of photosynthesis. Both processes need intact membranes that form intact compartments separated from the rest of the cell to function.

The respiratory chain can be characterised as the process during which NADH + H^+ and $FADH_2$ are oxidised under consumption of oxygen and water and ATP are generated. The generation of water (oxyhydrogen reaction) is strongly exergonic:

$$2\ H_2 + O_2 > 2\ H_2O.$$

In the respiratory chain, this energy is gradually set free in a number of intermediate steps (an electron transport chain) and can thus be used for the generation of energy-rich compounds.

— NADH + H^+ is oxidised resulting in NAD^+. The hydrogen is passed on to a further hydrogen-acceptor, the FAD.

— $FADH_2$ is oxidised by cytochromes. Cytochromes (cytochrome b, cytochrome c, cytochrome a, etc.) are proteins with haeme-groups (porphyrin rings) as cofactors. Located in the centre of this ring is an iron ion of changing valency:

$$Fe^{III} <> Fe^{II}$$

While FAD accepts both electrons and protons, cytochromes transport only electrons. The protons are given off into solution. In the respiratory chain, the FeIII of cytochrome b is at first changed into Fe^{II}. In plant cells, at least three different cytochrome b-types are in series.

The iron ion of cytochrome b achieves its original Fe^{III} state again by transferring one electron to cytochrome c which transfers it in a further step onto cytochrome a. A quinone/hydroquinone system may be inserted between the flavoprotein and the cytochromes.

— The reduced cytochrome a transfers the electrons onto oxygen that reacts immediately with free protons to form water.

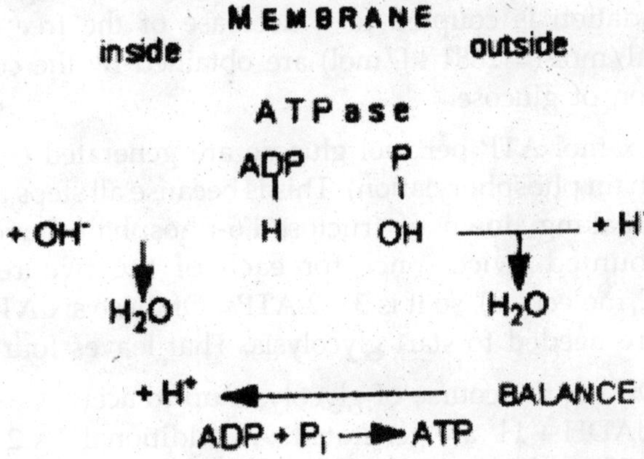

Electron Transport and Oxidative Phosphorylation

Each redox pair is characterised by a redox potential (E'°). This potential is proportional to the free energy (delta G°) of the reaction. Three redox pairs have enough difference in potential to drive the generation of one molecule of ATP each. Accordingly, three molecules of ATP, the equivalent of NADH + H^+ while a molecule of $FADH_2$ gives only two ATPs.

All the cofactors are bound to proteins and the proteins again are either integral components of the membrane or are closely associated. It is also important to know that all these proteins are oriented in the membrane in always the same way and that the membrane is impermeable for protons. Besides the mentioned proteins, one more enzyme is necessary, the

ATP synthase. How is the activity of the ATP synthase coupled to the electron transport chain? Proofs exists since quite some time already that this enzyme functions itself as a coupling factor.

A model of the way protons are translocated through the inner mitochondrial membrane. Please notice that the protons are not passed on but that OH-ions are set free at the outside of the membrane while protons are given away into the matrix at the inner surface. This results in a directed movement of the protons. The ATP synthase (also known as ATPase) has two neighbouring binding sites for ADP and phosphate.

We can now calculate the end result of glucose degradation: the oxidation is coupled to a decrease of the free energy; 686 kcal/mol (= 2881 kJ/mol) are obtained by the complete oxidation of glucose.

— Six mol ATP per mol glucose are generated (substrate chain phosphorylation). This is because all steps after the breaking down of fructose-l,6-phosphate have to be counted twice (once for each of the two .resulting C_3 molecules), so it is 3 x 2 ATPs. Of these six ATPs, two are needed to start glycolysis. That leaves four.

— During the course of glycolysis up to acetyl-CoA, 2 x 2 NADH + H^+ are generated. An additional 3 x 2 NADH + H^+ and 1 x 2 $FADH_2$ are produced in the citric acid cycle. One NADH + H+ gives three, one FADH2 two ATPs when fed into the respiratory chain. This sums up to 34 ATPs plus the 4 ATPs of glycolysis.

A total of 38 mol ATP are thus gained by the cell's degradation of one mol glucose. Since each energy-rich bond of ATP contains 7.3 kcal/mol (= -30.6 kJ/mol), the 38 ATP equal 277 kcal/mol (ca 1163 kJ/mol). This is 40.6 per cent of the theoretically possible gain. The other 59.4 per cent are set free as heat. This is a very high percentage compared to the gain of technical machines like steam or petrol engines that is around or below 20 per cent.

8

Energy Transfer

In many living organisms, including man, the *oxidative breakdown* of carbohydrates, fats and proteins is the principal source of energy for many activities of a cell. Oxidation originally means reaction with oxygen, but today, in a chemical context, it means *the loss of electrons by a molecule* or *atom*. Reduction, on the other hand, means the gain of electrons by the molecules or atoms. Oxidation of one substance must necessarily accompany the reduction of another. Hence, a single term *"redox reaction"* is used for any oxidation-reduction reaction. In any redox reaction, the number of electrons lost by the reducing agent must be equal to the number of electrons gained by the oxidising agent.

Role of Biological Oxidation

Biological oxidation means the oxidation of foods and substrates, that is carried out inside the living cells. In biological oxidation, the food molecules are broken down into smaller fragments and energy is released step by step. Although ordinary combustion or burning of fuels is similar in some way to the biological oxidation, yet there are certain fundamental differences as mentioned below:

1. In combustion, many chemical bonds are broken simultaneously releasing the energy with a great deal of heat. In biological oxidation, on the other hand, energy is released gradually by breaking bonds one after another and temperature is kept low.
2. Most of the energy of combustion is dissipated as heat, whereas the energy released in biological oxidation creates new chemical bonds and only a little energy is lost as heat. By placing the energy in small packages (chemical bonds) cellular metabolism can transfer energy to wherever a cell needs it.
3. Another important feature of biological oxidation is that it occurs in a series of many successive and coordinated enzymatic reactions and not in a single chemical step. Thus each of the many steps constituting biological oxidation is catalysed by a specific enzyme.
4. Oxygen is not directly necessary in biological oxidation. However, in the two processes of combustion and cellular oxidation, the potential energy within a given amount of an organic compound will always yield the same amount of energy, whether by one process or by the other. This is an accordance with the first law of thermodynamics.

The biological oxidation may take place in the following three ways:

(i) *By addition of oxygen*

$$C + O_2 \longrightarrow CO_2$$

(ii) *By removal of hydrogen*

$$\underset{\text{Lactic acid}}{CH_3CHOH.COOH} \longrightarrow \underset{\text{Pyruvic acid}}{CH_3.CO.COOH} + 2H$$

(iii) *By an increase in positive valency*

$$2Fe^{++} (Cl^-)_2 + Cl_2 \longrightarrow 2Fe^{+++} (Cl^-)_3$$

In this redox reaction, the ferrous ion is oxidised to the ferric state by losing an electron to chlorine.

In a simple form of biological oxidation, an enzyme *dehydrogenase* catalyses the transfer of two hydrogen atoms (or electrons) from a substrate to molecular oxygen. The hydrogen atoms unite with oxygen, which is reduced to form hydrogen peroxide. Since H_2O_2 is toxic, it is immediately converted into H_2O and O_2 by the enzyme *catalase*, which is present in all cells.

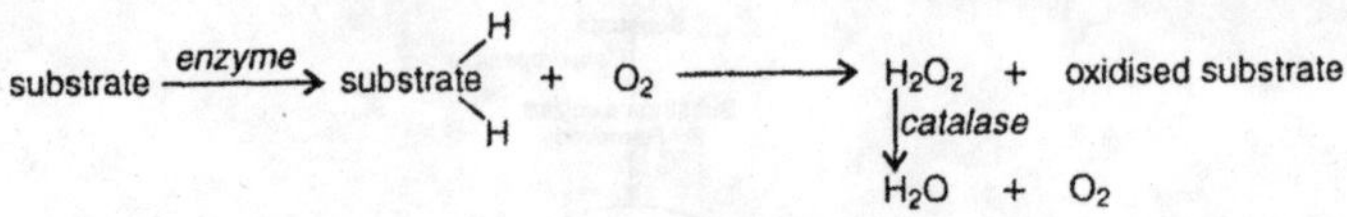

In the above reaction, the removal of activated hydrogen from the substrate is a simple process. But in most biological reactions, the process involves several intermediate steps.

All biological oxidations are actually *dehydrogenations, i.e.,* removal of hydrogen atoms and electrons. However, it is not easily possible to remove electrons because they cannot exist freely and as soon as they are removed they must be transferred to another molecule which becomes reduced.

Thus, all biological oxidations are oxidation-reduction reactions that are catalysed by the enzymes called *dehydrogenases.* In these reactions one molecule is every time oxidised and the other is reduced. In all living cells there are a number of organic molecules functioning as *hydrogen* or *electron acceptors.* They are called *co-enzymes.* The most important co-enzymes are nicotinamide adenine dinucleotide (NAD), nicotinamide adenine dinucleotide phosphate (NADP), flavin mononucleotide (FMN) and flavin adenine dinucleotide (FAD). These co-enzymes attach themselves to *dehydrogenases* only at the time of reaction and are present in the medium rest of the time. Every co-enzyme may function again and again as it is not used up in the course of functioning. This is because the hydrogen atoms or electrons transferred to them do not remain permanently attached but are transferred in the next reaction usually to another co-enzyme, and finally to the oxygen itself. The following reaction shows how NAD participates in the biological oxidation of lactic acid to pyruvic acid.

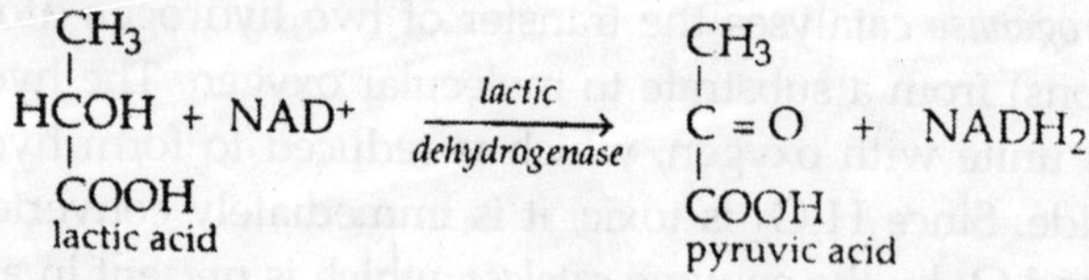

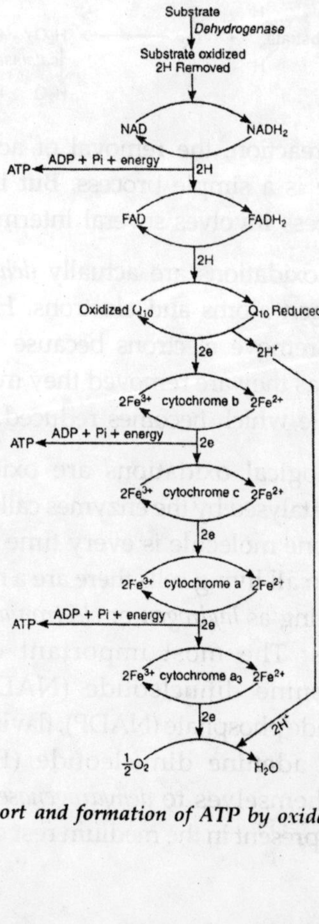

The electron transport and formation of ATP by oxidative phosphorylation.

Chains of Biological Oxidation

1. The first step in biological oxidation, consists of the activation of two hydrogen atoms and their electrons from a substrate by the enzyme dehydrogenase.
2. The hydrogen atoms and their electrons are then transferred to the first hydrogen acceptor, the NAD, which is then reduced.
3. From the reduced NAD or $NADH_2$ the hydrogen atoms and their electrons are passed on to the FAD, which is in turn reduced to $FADH_2$.
4. In the next step $FADH_2$ is oxidised when it delivers the hydrogen atoms and their electrons to the co-enzyme Q_{10}.
5. From reduced co-enzyme Q_{10} hydrogen atoms are released in the cytoplasm and electrons are now passed on to a series of cytochromes (b, c, a and a_3). In this process, each cytochrome is alternately reduced and oxidised.
6. In the last step, the electrons are finally transferred from cytochrome a_3 (also called cytochrome oxidase) to the final acceptor or the molecular O_2. Oxygen then unites with hydrogen atoms (protons) set free earlier in the cytoplasm to form water.

$$2\text{cytochrome } a_3\text{–}(Fe^{++}) + 2H^+ + 1/2\ O_2 \longrightarrow 2\text{cytochrome } a_3\text{–}(Fe^{+++}) + H_2O$$

The above electron transport chain is present in the mitochondria in which the biological oxidations are carried out. However, it may be mentioned that most of the hydrogen atoms and their electrons that are passed on to the electron transport chain, are removed by different dehydrogenases from different substrates of Kerbs cycle and only 2 pairs of hydrogen atoms are released in glycolysis.

Oxidative Phosphorylation and Formation of ATP: In

biological oxidation, the energy released in the oxidation of a substrate is used to create new ATP molecules. The formation of ATP is possible by *oxidative phosphorylation*, a process in which oxidation is coupled with phosphorylation and which occurs in the mitochondria of living cells.

The general plan of *oxidative phosphorylation* has been explained by *chemiosmotic* or *proton-motive* mechanism (hypothesis) proposed by P. *Mitchell* in 1961. The hypothesis proposes that the primary product of electron transport is the formation of an *electrochemical proton gradient* across the inner mitochondrial membrane.

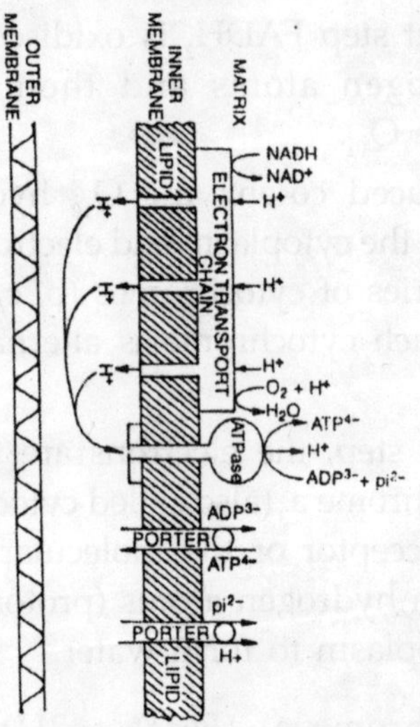

Oxidative phosphorylation by proton-motive mechanism.

ATP synthesis results from the spontaneous flow of protons down this gradient; the path leads through the membrane via a complex of proteins known as the *coupling factor*. The gradient is established by an obligatory movement of protons during respiration and is preserved by the presence of a *membrane with generally low permeability to protons*, except at special locations where the energy of the proton gradient is harnessed to do cellular work. The critical proton barrier is the inner mitochondrial membrane. Exactly how the electro-chemical gradient of protons, *the proton-motive force* (pmf), as *Mitchell* first called it, is used to phosphorylate ADP remains a central mystery of energy metabolism. The following reaction is derived by the proton current.

$$ADP^{3-} + Pi^{-} + H^{+} \longrightarrow ATP^{4-} + H_2O$$

The source of protons delivered to the outside of the membrane is either protons generated as certain components (*e.g.*, NADH, $FMNH_2$) of the electron transport chain are oxidised, or through a vectorially arranged process in which various electron carriers, in their reduced states, would dissociate one or more protons, releasing them to the outside of the membrane and recapture protons from the matrix when reoxidised. There is secretion of $4H^+$ per electron pair at each phosphorylating site. This is sufficient to synthesise ATP and to titrate Pi as it is transported into the matrix. When the protons have passed through the ATPase and return to the matrix, they have completed a proton circuit. This flow of protons from high protic potential to low protic potential is analogous to an electric circuit *(Smith* et al, 1983).

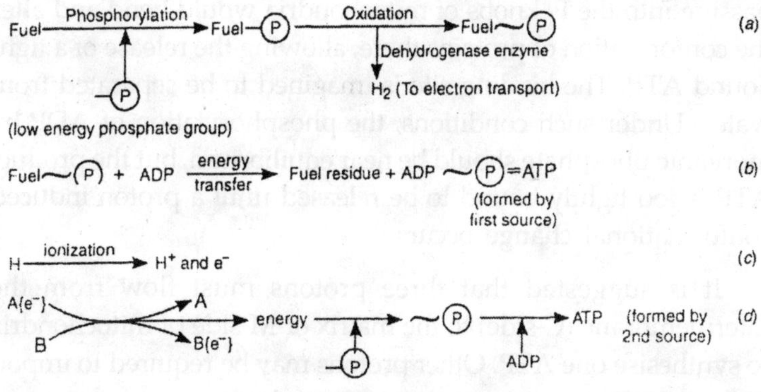

Fig. Showing phosphorylation, oxidation (dehydrogenation) and two sources of ATP formation (one being the oxidation of phosphorylated food by dehydrogenation and the other being the dehydrogenation energy or the creation of high-energy phosphate bond during electron transfer).

Another source of ATP is dehydrogenation energy, the energy liberated in the transfer of electrons. This energy is also used in the generation of more ATP molecules by oxidative phosphorylation. It has been stated that NAD works as the first hydrogen acceptor. For every molecule of $NADH_2$ that is oxidised via the electron transport system, 3 molecules of ATP are formed.

$$NADH_2 + 1/2O_2 + 3ADP + 3Pi \longrightarrow NAD + 3ATP + H_2O$$

Of the 3 ATP molecules, one is formed during the passage of hydrogen atoms from $NADH_2$ to FAD, and the rest 2 are formed during the transfer of electrons from co-enzyme Q_{10} to molecular O_2.

Thus, the most significant product of biological oxidation is ATP and its formation depends upon the coupling of oxidation with phosphorylation. The possible mechanism of oxidative phosphorylation is the *conformational coupling hypothesis* put forward by P. *Boyer* (1977). The hypothesis proposes that protons passing into the F_1 knobs of mitochondria would bend and alter the conformation of proteins there, allowing the release of a tight bound ATP. The binding site is imagined to be separated from water. Under such conditions, the phosphorylation of ADP by inorganic phosphate should be near equilibrium, but the product ATP is too tightly bound to be released until a proton induced conformational change occurs.

It is suggested that three protons must flow from the intermembrane (C-side) to the matrix or M-side of mitochondria to synthesise one ATP. Other protons may be required to import Pi and ADP or to export the ATP product.

Transfer of Energy in Cellular Metabolism: ATP plays the most important role in the transfer of energy. Besides ATP a number of compounds are used in energy transfer in cells.

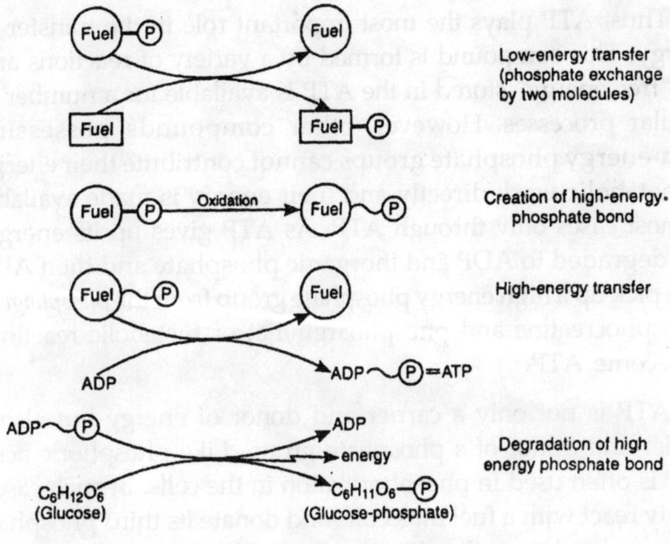

The four ways in which the energy of the phosphate bond is transferred.

Low-energy Transfer: It means exchange of phosphate by two molecules. The most important high-energy bond is the *phosphate bond, i.e.,* the bond which joins a molecule to a phosphate group (phosphorylation). Once an organic molecule is so phosphorylated, its phosphate group may then be transferred in turn to various other organic compounds. This is called *low energy transfer, i.e.,* the low energy value of the phosphate bond does not change as in phosphate exchange by two molecules.

High-energy Transfer: In many cases the transfer of energy occurs between organic molecules possessing the high-energy phosphate bond and the ADP/ATP system which harvests respiratory energy. This is called *high-energy transfer.* For example, when a fuel molecule is phosphorylated and oxidised, it may possess a high-energy phosphate bond. If the latter is then transferred to ADP, the ADP becomes ATP by accepting a high energy phosphate bond from the fuel. Through this type of transfer, some of the energy of a fuel molecule becomes incorporated into ATP.

Thus, ATP plays the most important role in the transfer of energy. This compound is formed by a variety of reactions and the "free energy" stored in the ATP is available for a number of cellular processes. However, other compounds possessing high-energy phosphate groups cannot contribute their energy to metabolic work directly and their energy is made available in most cases only through ATP. As ATP gives up its energy, it is degraded to ADP and inorganic phosphate and then ADP may pick up a high energy phosphate group from the *phosphagens* (phosphocreatine and phosphoarginine) or metabolic reactions to become ATP.

ATP is not only a carrier and donor of energy but also a carrier and donor of a phosphate group. Like phosphoric acid, ATP is often used in phosphorylation in the cells. In such cases, it may react with a fuel molecule, and donate its third phosphate to the molecule, *e.g.*, the phosphorylation of glucose. To complete such reactions, only a little energy of ATP is used and the rest is dissipated as heat.

Besides ATP, several other compounds (*e.g.*, co-enzyme A, $NADH_2$, and phosphagens) also play a role in energy transfers in cellular metabolism.

9

Significance of Hormones

Hormones are chemical messengers that coordinate the activities of different cells in multi-cellular organisms. The term *hormone* (from Greek, meaning to spur on) was first used in 1904 by William Bayliss and Ernest Starling to describe the action of secretin a molecule secreted by the duodenum that stimulates the flow of pancreatic juice. Some useful concepts emerged from their work were as follows:

1. Hormones are molecules synthesised by specific tissue, the *glands*.
2. They are secreted directly into the blood which carries them to their sites of action.
3. They specifically alter the activities of responsive tissues which are called target organs or *target cells*.

Categories of Hormones

Hormones are chemically diverse. We can classify them into 4 categories on the basis of their chemical structure:

1. *Small Molecules Derived from Amino Acids:* epinephrine, thyroxine.

2. *Polypeptides* or *Proteins:* oxytocin, insulin, thyroid-stimulating hormone and many growth factors.
3. *Steroid hormones:* made from cholesterol, *e.g.*, cortisol.
4. *Eicosanoids:* formed from arachidonate, a C_{20}-poliunsaturated fatty acid, *e.g.*, prostaglandin PGA,.

Role of Cyclic Amp as a Second Messenger in Hormone Action:

Discovery of Cyclic AMP: Earl Sutheland, Leon Heppel, and David Lipkin working independently in 1950s discovered a new nucleotide adenosine 3', 5'-monophosphate, now commonly referred to as *cyclic AMP* or cAMP.

Synthesis of Cyclic AMP: Cyclic AMP is formed from ATP by the action of *adenylate cyclase,* an integral membrane protein.

$$\text{ATH} \xrightarrow{Mg^{2+}} \text{cyclic AMP} + \text{PPi} + H^+$$

This reaction is slightly endergonic having a AG°' of about 1.6 kcal/mol. It is driven in the direction of the synthesis of cyclic AMP by the subsequent hydrolysis of *pyrophosphate.*

Degradation of Cyclic AMP:

Cyclic AMP is degraded by a specific *phosphodiesterase,* which hydrolyse it to AMP.

$$\text{Cyclic AMP} + H_2O \xrightarrow{Mg^{2+}} \text{AMP} + H^+$$

This reaction is highly exergonic having a $\Delta G°'$ of about— 12 kcal/mol. yet cyclic AMP is a very stable compound in the absence of the phosphodiesterase.

Cyclic AMP is a Second Messenger in the Action of Many Hormones: Sutherland's work led to the concept that cyclic AMP is a *second messenger* in the action of some hormones. The first messenger is the hormone itself. The essential features of this concept are:

1. The plasma membrane contains receptors for hormones.

2. The combination of a hormone with its specific receptor leads to the activation of a *denylate cyclase,* which is also bound to the plasma membrane.
3. The increased activity of adenylate cyclase increases the level of cyclic AMP in the cytosol.
4. Cyclic AMP then acts inside the cell to alter the rate of one or more processes.
5. *The most important feature of the second messenger model is that the hormone need not enter the cell.* Its impact is made at the cell membrane. The biological effects of the hormone are mediated inside *the* cell by cyclic AMP rather than by the hormone itself.

The second messenger model makes several predictions that can be experimentally tested:

1. Adenylate cyclase in a target cell is stimulated by hormones affecting that cell. Cells that do not show a characteristic biological response to a particular hormone do not increase their cyclase level in its presence.
2. The change in concentration of cyclic AMP in a target cell occurs at the same time as the response to hormonal stimulation. Variations in hormone levels match by variations in the concentration of cyclic AMP.
3. Inhibitors of the phosphodiesterase such as theophylline and caffeine act synergistically with hormones that use cyclic AMP as a second messenger.
4. The biological effect of a hormone are mimicked by the addition of cyclic AMP or a related compound to the target cells.

Experiments based on these criteria have proved that *cyclic AMP is a second messenger for many hormones in addition to epinephrine and glucagon.* These are as follows:

Hormones using cyclic AMP as a second messenger

1. Calcitonin

2. Chorionic gonadotropin
3. Corticotropin
4. Epinephrine
5. Follicle-stimulating hormone
6. Glucagon
7. Luteinising hormone
8. Lipotropin
9. Melanocyte-stimulating hormone
10. Norepinephrine
11. Parathyroid hormone
12. Thyroid stimulating hormone
13. Vasopressin.

Mechanism of second messenger concept of cyclic AMP

1. *A guanyl-nuceotide-binding protein is an intemediary in the activation of adenylate cyclase.* This signal coupling protein was named G *protein* (G).
2. The hormone-receptor complex does not directly stimulates adenylate cyclase. Rather, the activated receptor stimulates the G protein, which carries the excitation signal to adenylate cyclase. This signal-transduction process takes place in the plasma membrane.
3. Alfred Gilmans has shown that the G protein is a peripheral membrane protein consisting of α β and γ subunits. It interconverts between a GDP form and a GTP form. The GTP form activates adenylate cyclase, whereas the GDP form does not in the absence of hormone, nearly all of the G protein is in the inactive GDP form.

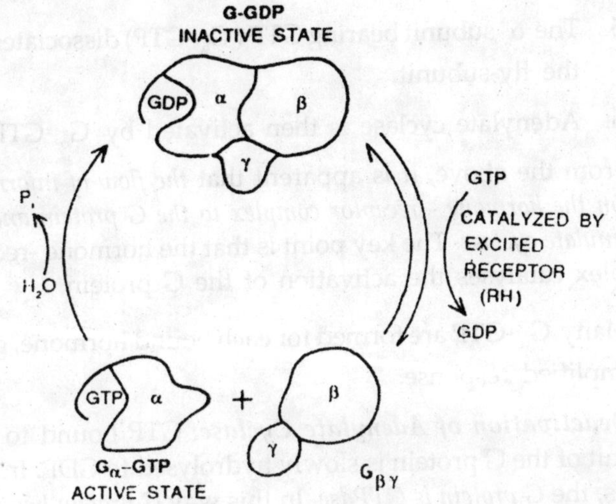

G protein inter-converts between an inactive GDP form and an active GTP form. The exchange of GTP for bound GDP is catalysed mediated by a hormone-receptor complex.

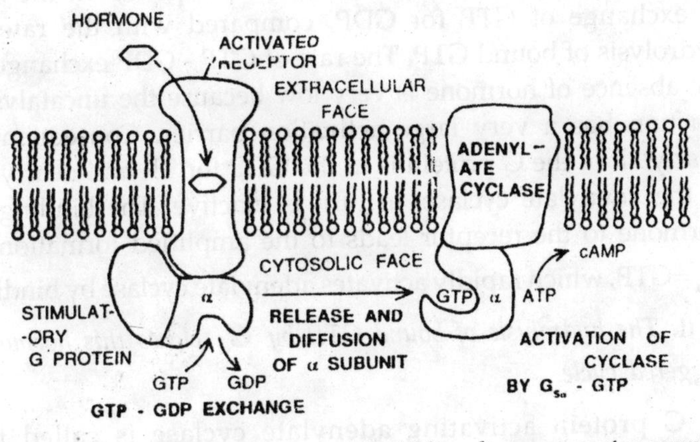

The activation of adenylaic eyclase mediated by a hormonereceptor complex.

4. *The binding of hormone to the receptor triggers the exchange of GTP for bound GDP.* In this process, the hormone-receptor complex (i) binds to the G protein, (ii) induces the release of bound GDP, and (iii) allows GTP to enter.

5. The α subunit bearing GTP (G_α -GTP) dissociates from the fly-subunit.

6. Adenylate cyclase is then activated by G_α -GTP.

From the above, it is apparent that *the flow of information is from the hormone - receptor complex to the G protein and then to adenylate cyclase.* The key point is that the hormone -receptor complex catalyses the activation of the G protein.

Many G_α -GTP are formed for each bound hormone, giving an amplified response.

Deactivation of Adenylate Cyclase: GTP bound to the a subunit of the G protein is slowly hydrolysed to GDP. In other words, the *G protein is GTPdse.* In this way *G protein has a built in device for switching off.*

The proportion of G protein in the GTP state, and hence of adenylate cyclase in the active form, depends on the rate of exchange of GTP for GDP, compared with the rate of hydrolysis of bound GTP. The rate of GTP - GDP exchange in the absence of hormone is very low because the uncatalysed reaction has a very large activation barrier. Consequently, nearly all of the G protein is in the GDP for TR and nearly all of the adenylate cyclase, in turn is inactive. The binding of hormone to the receptor leads to the amplified formation of G_α =GTP, which rapidly activates adenylate cyclase by binding to it. *The hydrolysis of bound GTP by G_α closes this hormone-triggered cycle.*

G protein activating adenylate cyclase is called the *stimulatory G protein* (Gs).

7. The activated adenylate cyclase in the plasma membrane catalyses the formation of cyclic AMP from ATP.

$$\text{ATP} \xrightarrow[\text{Mg}^2]{\text{Adenylate cyclase}} \text{Cyclic AMP} + \text{PPi} + \text{H}^+$$

8. The increased intercellular level of cyclic AMP activates a *protein kinase.* This kinase is inactive in the absence of cyclic AMP. The binding of cyclic AMP allosterically stimulates the protein kinase.
9. In skeletal muscle, this protein kinase phosphorylate both *glycogen synthase* (rendering it inactive) and *phosphorylase kinase* (rendering it active). In this way, cyclic AMP stimulates glycogen break down and stops glycogen synthesis. A similar mechanism operates in the liver.

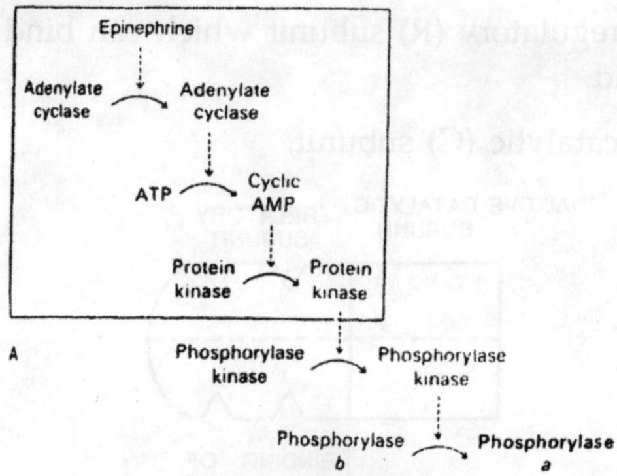

A. Glycogen degradation

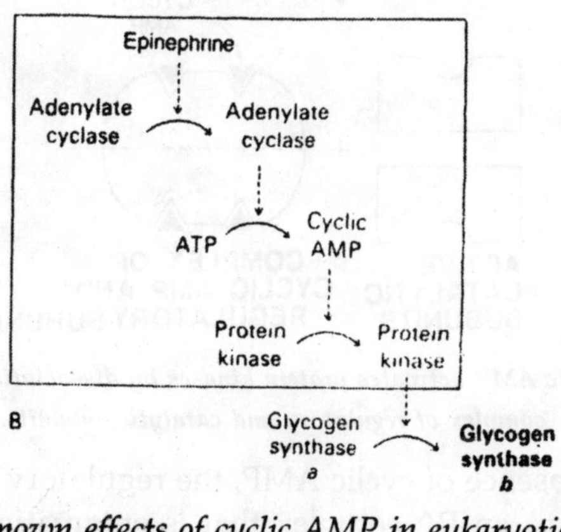

10. All *known effects of cyclic AMP in eukaryotic cells results from the activation of protein kinases.* These protein kinases, which are activated by cyclic AMP

concentrations of the order of 10^{-8}m. modulate the activities of target proteins by phosphorylating seine residues in Arg-Arg-X-Ser-X sequences (X denotes any residue). Phosphorylation of target proteins generally changes their conformation and, in enzymes, increases or decreases their activity, thus producing changes in metabolism.

Mechanism of activation of the cyclic-AMP controlled kinase.

The protein kinase in muscle consists of two kinds of subunits:

(i) a regulatory (R) subunit which can bind cyclic AMP and.

(ii) a catalytic (C) subunit,

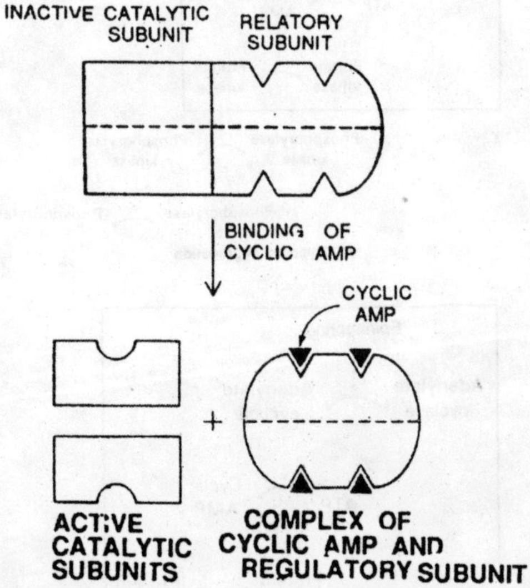

Cyclic AMP activates protein kinases by dissociating the complex of regulatory and catalytic subunits.

In the absence of cyclic AMP, the regulatory and catalytic subunits form an R^ complex that is enigmatically inactive. The binding of two molecules of cyclic AMP to each of the regulatory subunits leads to the dissociation of the R^ complex into an R, subunit and two C subunits. These free catalytic subunits are then enzymatically active. It is inferred, therefore, that *the binding of cyclic AMP to the regulatory subunit relieves its inhibition of the catalytic subunit. Cyclic AMP acts as an allosteric effector.*

The adenylate cyclase cascade amplifies the hormonal signal

The concentration of many hormones in the blood is of the order of 10^{-10}m. The adenylate cyclase cascade greatly amplifies such weak hormonal signals. Three steps in the cascade provide a high degree of amplification:

1. Each hormone-receptor complex catalyses the formation of many G_α-GTP.
2. Many molecules of cyclic AMP are formed by an activated adenylate cyclase
3. Each cAMP-activated protein kinase can alter the activity of many molecules of each target protein

Cyclic AMP affects a very wide range of cellular processes.

For example:

1. It enhances the degradation of storage fuels.
2. It increases the secretion of acid by the gastric mucose.
3. It leads to the dispersion of melanin pigment granules.
4. It diminishes the aggregation of blood platelets.

Catalytic Substances in the Body

Besides hormones, a number of other hormone-like substances are also produced. These are *(A)* prostaglandins, (B) pheromones, (C) neurohormones, (D) neurotransmitters, and *(E)* neuropeptides. However, they differ from hormones in several ways and their effects are wide ranging.

Prostaglandins: Prostaglandins (PG) are a group of biologically active lipids which appear to play important roles in many reproductive processes. These substances were first found in human seminal fluid and the name 'prostaglandins' was given to them by *von-Euler* (1933-34). Actually the term was a misnomer because the main source of prostaglandins in semen is the seminal vesicles and not the prostate gland. Since then prostaglandins have also been found in the semen of other vertebrates like sheep, monkey and goat. In human beings they occur in other body fluids also including the menstrual fluid of women, and in tissues of pancreas, stomach, lung, kidney, brain, spinal cord, iris, and placenta, etc.

Prostaglandins differ from all other hormones in being fatty acids (that is, incorporating the COOH group) and all those so far examined contain 20 carbon atoms. The primary prostaglandins (PGE and PGF) are synthesised in the body from some polyunsaturated fatty acids by the formation of a five-member ring and the incorporation of 3 oxygen atoms at certain positions. Thus, all prostaglandins have the same basic "skeleton", *prostanoic acid* and are derived from essential fatty acids by cyclisation and oxidation. There are two main sets of prostaglandins, the E

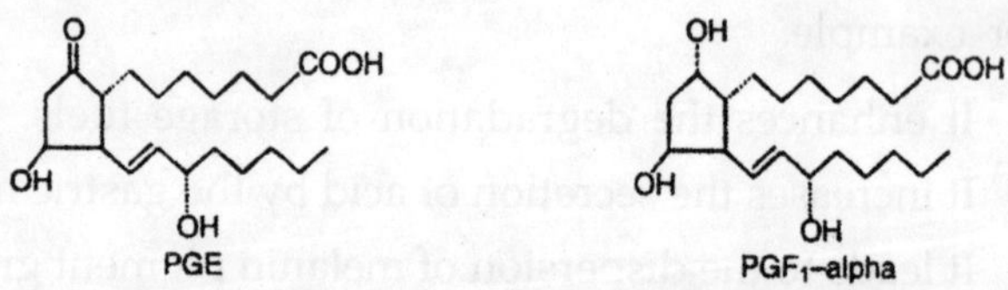

The structure of PGE_1 and PGF_1 prostaglandins.

and F series, each having three members E_1, E_2 and E_3 and $F_1\alpha$, $F_2\alpha$, and $F_3\alpha$. The two series differ from each other in that the PGE forms contain a keto oxygen at C-9 and a hydroxyl group at C-ll, whereas the PGF structures have a hydroxyl group at both positions in the cyclopentane ring. Prostaglandins of the 1, 2 and 3 series possess 1, 2, and 3 double bonds respectively. The primary structures can thereafter be converted into other members (PGA, PGB) of the prostaglandin family.

Biological Role of Prostaglandins: Prostaglandins interact with specific receptors that are localised in the plasma membranes of cells. The receptors for the PGE forms are different from those for the PGF. Prostaglandins are among the most potent of all the known biological substances producing marked effects in extremely small doses. In general, their effects are based on the regulation of the activity of smooth muscles, of secretion (including some endocrine gland secretions), or of blood flow. It seems possible that prostaglandins show their effects by affecting adenylcyclase and cyclic AMP. Through these actions prostaglandins are capable of affecting many aspects of human physiology, and this accounts for their importance in pharmacology.

According to *Pike* (1971), prostaglandins may be formed by the lungs during anaphylaxis, by the kidney when its blood supply is restricted, by the skin during human allergic contact eczema and certain experimental inflammatory conditions. This suggests that they may play an important role not only in normal physiological activities but also in certain pathological conditions. The wide range of their physiological effects also includes blood pressure regulation, lipid and carbohydrate metabolism, and tissue inflammation in response to disease or wounds.

Effects on Female Reproductive System: The effects of prostaglandins on female reproductive system are very striking. *Bygdeman* (1971) has shown that an intravenous injection of a very low dose of either PGE_2 or PGF_2 alpha stimulates the contration of the uterus and can be used to initiate labour in women or to produce abortion. Drs. *Sune Bergstom* and *Bengt Samuelson* have won the 1982 Nobel prize with *Dr. John Vane* for the development of a home abortion treatment based on prostaglandin at Stockholm's Karolinska Institute. The technique is an abortion suppository which involves the prostaglandin's chemical action causing cramping of the uterine muscles and the abortion of the foetus. However, the technique has been criticised by Dr. *P.H. Petersen.* Now there seems to be a strong possibility that prostaglandins or their synthetic analogues may be used as important agents to control population.

Effect on Blood Flow Control and Immune Responses: Prostaglandins are released into the local tissue fluids and the circulating blood under both physiological and pathological conditions. Although some of them cause vasoconstriction, most of men seem to be mainly vasodilator agents. Administration of PGE_1 or PGE_2 intravenously augments blood flow by inducing a generalised vasodilation, with decreased peripheral resistance. The inhibitory actions of prostaglandins or immune responses are due to inhibition by PGE_1 of antigen activation of thymus-dependent lymphocytes, a process of importance in immunologic responses.

Effect on Intestinal Secretions: The prostaglandins inhibit gastric secretion but stimulate pancreatic and mucous secretion in the intestine and markedly increase intestinal motility. *Andre Robert's* experiments on rats and dogs have shown that the stomach normally produces E prostaglandins which serve to regulate gastric secretion under normal circumstances and thus protect the stomach against ulceration. If his studies are confirmed in man, the administration of E prostaglandins may be helpful in treating ulcer patients who lack this normal protection. Prostaglandins also appear to enhance ion flux across epithelial membranes.

Effect on Nerve Fibres: Recently it has been shown that production and release of prostaglandins can be evoked by stimulation of nerves and there exists an interrelation between the prostaglandins and the substances involved in the transmission of nerve impulses. Normally, prostaglandins play a negative feedback role on the release of noradrenaline by sympathetic nerve fibres. They also act as modulators of the actions of chemical transmitters released from nerve endings and of hormones from endocrine glands.

Effect on inflammation: Among the many tissue products responsible for tissue changes during inflammation, prostaglandins are considered important. Prostaglandins produced at a local site as a result of injury, burns, etc., act as mediators of the inflammatory response.

Effect on Endocrine Glands: It has been also suggested that prostaglandins act as controlling agents which modify the action of hormones on target cells. Several endocrine glands such as the thyroid, adrenal cortex, ovary, and parathyroids respond to prostaglandin administration with increased secretion of their hormones. In addition, through their effects on adenylate cyclase the prostaglandins may modulate hormonal actions. The effects of many hormones that increase the activity of the adenyl cyclase system are antagonised or enhanced by PGE_1 or PGE_2. For example, LH, by mobilising triglycerides and phospholipids in the ovary, might increase the synthesis of prostaglandins through a greater concentration of their precursor unsaturated fatty acids. The prostaglandins so synthesised could modify the metabolic consequences of LH on the ovarian cell.

Effect on Reproduction: The secretory epithelium of seminal vesicles secretes a mucoid material containing fructose, citric acid and other nutrient substances as well as large quantities of prostaglandins and fibrinogen. The prostaglandins are believed to aid fertilization in two ways. First, by reacting with the cervical mucus to make it more receptive to sperm movement. Second, by possibly causing reverse peristaltic contractions in the uterus and fallopian tubes to move the sperm towards the ovaries.

Pheromones: The term *'pheromone'* was originally applied to the sex *attractants* of insects but now it includes all kinds of chemical substances released into the environment to aid in the integration of behaviour among members of the same animal species. According to *Karlson* (1960), chemical signals exchanged between individuals of the same species are pheromones, which cause a specific reaction either in the form of a behavioural response or in a particular developmental process.

The pheromones differ from hormones in several respects:

(a) they are produced by the exocrine glands,

(b) they are transmitted via the external environment,

(c) they are typically more species-specific than the hormones, and

(d) they evoke specific behavioural, developmental, or reproductive responses in the bodies of other individuals of the same animal species, whereas the hormones regulate organism's own physiology.

Thus, pheromones are not hormones but the activity of the glands that produce them often depends on hormonal stimulation. The pheromones act, in general, on the hypothalamus via olfactory pathways and thereby alter the behaviour or the physiology of the recipients. The source of such substances, also called ectohormones, are a variety of glands found in the skin or other places in insects and mammals.

On the basis of their mode of action pheromones are grouped into two categories. These are:

Signalling or Releaser Pheromones: The pheromones that produce a simple "releaser" effect, *i.e.*, a single, rapid, specific response mediated directly by the central nervous system, are called *releaser pheromones*. To this category belong the sex attractants (sex pheromones), and the trail and alarm substances. The ant *Iridomyrmex pruinosus* uses 2-heptanone as an alarm pheromone. Among insects the sex attractants have been identified in silk moth, gypsy moth, cockroach, honey bee and mosquitoes. Female silkworm moths *(Bombyx mori)* secrete a sexual attractant, *bombykol:*

$$CH_3—(CH_2)_3—CH = CH —CH = CH—(CH_2)_8—CH_2OH$$

Only the antennal chemoreceptors of the male are excited by this substance, those of the female are insensitive. When sensed by the male, a behaviour pattern is initiated such that the male flies to the female.

Many studies have been made on the sexual attractants of the gypsy moth *(Porthetria dispar)* whose females secrete the pheromone *glyplure:*

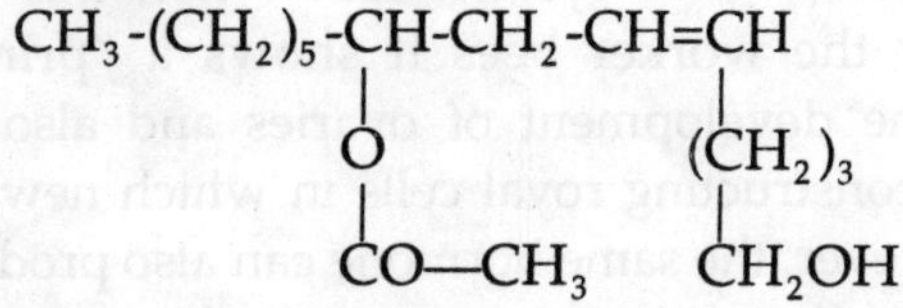

Depending on the wind speeds, gypsy moth females can attract males from up to 4500 m away.

With respect to communication, pheromones cause a variety of actions on the part of individuals. They may cause the dispersal or aggregation of a population. When populations of flour beetle *Tribolium confusum* become dense, *quinones* released from the abdominal glands cause the individuals to disperse. Insect aggregation, through the action of pheromones, is exemplified by the behaviour of honey bees and other social insects *(Renner,* 1960; *von Frisch,* 1967).

When worker honey bees find a good source of food, they release pheromones from an abdominal scent gland to aid other workers in finding the food. Similar activity is found when a new hive is started. One component of the pheromone secretion is *geraniol,* but at least four other compounds are present *(Bock* and *Schearer,* 1962).

Primer Pheromones: The pheromones which trigger a chain of physiological events, or those that activate a long series of neuroendocrine events that develop slowly and require prolonged stimulation, are called *primer pheromones.* The primer pheromones often act by physiological inhibition but they can produce the opposite effect. For example, the adult males of the migratory locust secrete a volatile substance from their skin that accelerates the growth of young locusts and thus plays an important role in the formation of migratory locust swarms.

An important example of primer pheromones is the secretion produced by the mandibular glands of the honey bee queen. This substance has been recently identified as *9-ketodecanoic acid* and synthesised. When this pheromone ingested by the worker bees it shows a "primer" effect, inhibiting the development of ovaries and also preventing them from constructing royal cells in which new queens are reared. However, the same hormone can also produce releaser or signalling effects, since it serves as a sex attractant during the queen's nuptial flight.

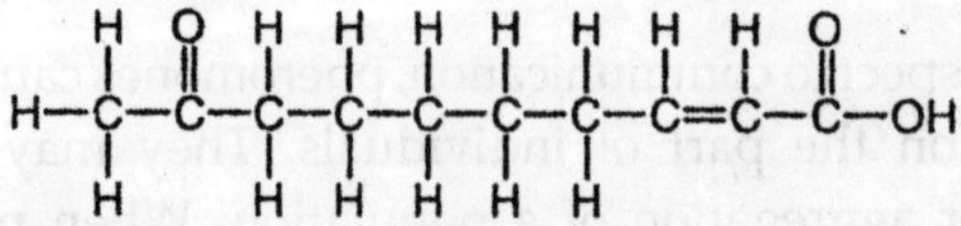

Honey bee queen substance (9-ketodecanoic acid).

Primer pheromones and sex attractants have been identified in a wide variety of terrestrial animals, including mammals. The sex attractants of civet *(Viverra zibetha)* and muskdeer *(Moschus moschiferus)* are called *civetone* and *muskone* respectively. These high molecular weight compounds contain about 16-17 carbon atoms and are highly volatile and diffusible. The civetone is secreted from the para-anal glandular pouch, and is used for defence and also a territory marker *(Wilson* and *Bossert,* 1963).

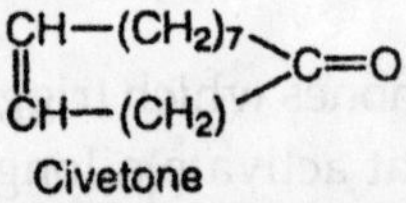

Civetone

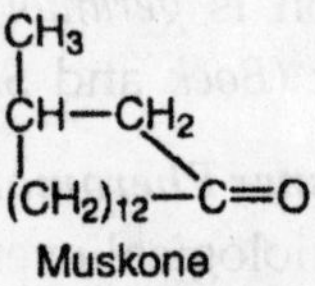

Muskone

The muskone functions as a sexual attractant. Another example of a mammalian pheromone is the one produced by male mice. *Helen Bruce* has described that the odour of a strange male of a different strain blocks the pregnancy of a newly impregnated female mouse *(Bruce effect).*

Pheromones in Humans: Now it is known that humans also produce sex pheromones. *Le Magnen* has reported that the odour of *Exaltolide* (a synthetic analogue of mammalian sex athractants, 14-hydroxy tetradecanoic acid) is perceived only by the sexually mature females.

The main source of human sex pheromones are some odoriferous glands in the skin, including the mammary glands, axillary glands, circumoral glands, circumanal glands, eyelid glands, glands of the external auditory canal and the glands of public region, etc. Saliva, urine, tears and vaginal secretions are other probable sources of human pheromones. It appears that pheromones also operate in the regulation of sexual behaviour in primates.

Neurohormones: The chemical messengers released by neurosecretory cells are called neurohormones. The neurohormones of the vertebrate neurohypophysis are known as *oxytocin* and *vasopressin*. *Harris* (1966) has suggested that the neurohormones are released by the neurosecretory cells in the hypothalamus and conveyed to the anterior pituitary by portal vessels where they regulate the output of hormones. Seven different neurohormonal substances that are peptides have been defined. They promote the release of thyrotrophin, ACTH, growth hormone, FSH and LH respectively, and inhibit the release of prolactin and melanocyte-stimulating hormone from the adenohypophysis.

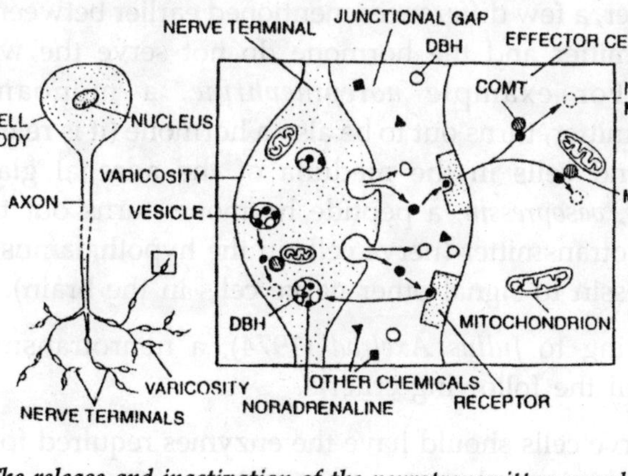

The release and inactivation of the neurotransmitter, noradrenaline.

Thus, neurohormones may perform regulatory functions directly by employing non-endocrine targets such as the contractile elements of the mammary glands, uterine muscle or kidney tubules, or indirectly by activating the endocrine glands, which in turn bring about adjustments by their own hormones.

Neurotransmitters: A neurotransmitter (neurohumour) is a chemical substance that is discharged from a nerve fibre ending, and then it reaches and is recognised by a receptor on the surface of a post-synaptic nerve cell (or effector cell) and either stimulates or inhibits the second cell. In many instances the neurotransmitter is a *monoamine* (a substance the nerve cell synthesises by making minor changes in an amino acid). Sometimes the neurotransmitter is the amino acid itself.

In any case its release and reception takes milliseconds, whereas the release of the hormone, its circulation in the blood and its influence on the target cells takes more time (minutes or hours). Thus, neurohumours are chemical messengers, but they differ from ordinary hormones because they do not depend on blood for their circulation. They also differ from neurohormones with respect to their source ; they are the product of the conventional nerve cells instead of coming from the neurosecretory cells.

However, a few differences mentioned earlier between the neurotransmitter and the hormone do not serve the whole purpose. For example *norepinephrine,* a monoamine neurotransmitter, turns out to be also a hormone (it is released by the gland cells in the medulla of the adrenal gland). Conversely, *vasopressin,* a peptide hormone, turns out to be also a neurotransmitter (nerve cells in the hypothalamus rely on vasopressin to signal other nerve cells in the brain).

According to *Julius Axelrod* (1974), a neurotransmitter should fulfil the following criteria.

(a) Nerve cells should have the enzymes required for the synthesis of the neurotransmitter.

(b) On stimulation nerve cells should liberate the neurotransmitter which should then react with a specific receptor on the post-junctional cell and produce a response.

(c) There should be a mechanism to terminate the action of the neurotransmitter rapidly.

On the basis of these criteria *acetylcholine* and *noradrenaline* (norepinephrine) have been established as the neurotransmitters and the nerve fibres that produce them are called "cholinergic" and "noradrenergic" respectively. Of the neurotransmitters secreted by enteric neurons, acetylcholine most often excites gastrointestinal activity; norepinephrine, on the other hand, almost always inhibits gastrointestinal activity. However, some other chemicals fulfil the above criteria only partly. They are dopamine, adrenaline, serotonin, etc.

Neurohumors are involved in a variety of physiological processes, some of which are given below:

(i) They make the heart beat faster (sympathin or norepinephrine) or slower (acetylcholine).

(ii) They are responsible for synaptic transmission and neuromuscular transmission.

(iii) They cause the glands to synthesise hormone-producing enzymes or to secrete the hormones.

(iv) Through these chemicals the brain regulates movement and changes in the mood and behaviour.

(v) Perhaps they also play a role in the diseases such as depression, schizophrenia, motor disorders, and hypertension, etc.

Neuropeptides: Recently *Bloom* (1981) has described that the *neuropeptides* are short chains of amino acids that are active in the nervous system. In some cases they transmit signals between nerve cells and also serve the body as a hormone. They differ from small-molecule, rapidly acting transmitters (acetylcholine, etc.) in their mode of action. They usually cause more prolonged actions; for example hypothalamic-releasing hormones (LH-releasing hormone, thyrotropin-releasing hormone, somatostatin) and pituitary peptides (ACTH, oxytocin, vasopressin, etc.). Some neuropeptides that act on gut and brain are *Leucine enkephalin, neurotensin,* etc.

Bloom (1981) has described that the nerve cells containing *endorphin* are found in the rat in the arcuate nucleus of the hypothalamus. The axons of the cells project along pathways in the brain similar to those of the cell containing oxytocin.

However, the nerve cells containing enkephalin are widespread in the brain of the rat.

Bibliography

Achaya, K.T. : *Indian Dairy Products,* Asia Publishing House, Bangalore, 1974.

Adam, H. : *Food Science,* John Wiley and Sons, New York, 1982.

Addis, Ababa : *Education and Training in Science and Technology in Asia,* UN Economic and Social Council, New York, 1989.

Arbuckle, S. : *Learning of Education,* Allyn and Bacon, Boston, 1965.

Bachman, W. : *Food Chemistry,* Maclaren and Sons Ltd., London, 1955.

Balsdon, J.P. : *Life and Leisure in Ancient Rome,* Bodley Head, London, 1966.

Banerjee, N. : *Indian Females in a Changing Industrial Scenario,* Indo-Dutch Series on Development Alternatives, Sage Publications, New Delhi, 1991.

Cain, M.L. : *Women and Technological Change in Development,* Westview Press, Colorado, 1981.

Carlton, E. : *Foundations of Technical Education,* Prentice Hall, New Jersey, 1964.

Carr, M. : *Appropriate Technology: Two Essays,* Intermediate Technology Development Group, London, 1982.

Crites, J.C. : *Vocational Psychology,* McGraw Hill, New York, 1969.

Donald, E. : *Learning of Technical Education,* Harper Row, New York, 1957.

Dumazedier, J. : *Towards a Society of Leisure,* Free Press, New York, 1967.

Duncan, W.A. : *Engendering School Learning: Science, Attitudes and Achievement among Girls and Boys in Botswana*, University of Stockholm, Stockholm, 1989.

Dyke, S .F. : *The Chemistry of Vitamins*, Interscience Publishers, New York, 1965.

Eley, G. : *Wild Fruits and Nuts*, E.P. Publishing Ltd., Yorkshire, 1976.

Franklin, U.M. : *Will Women Change Technology, or will Technology Change Women?*, The Canadian Research Institute for the Advancement of Women (CRIAW), London, 1985.

Gill, L.S. : *Science, Technology and Gender*, Polity Press in Association with the Open University, Cambridge, London, 1992.

Hanson, J.C. : *Vocational Guidance and Career Development*, MacMillan, New York, 1968.

Helen, Appleton : *Do it Herself: Women and Technical Innovation*, IT Publications, Mumbai, 1995.

Hoyt, K. : *Techniques of Education*, Science Research Associates, New York, 1959.

Hudson, F. : *Development in Food Proteins*, Applied Science Publisher, London, 1982.

James, A.R. : *Education : A System for the Seventies*, Charles E. Marrill Pub-lishing Co., Columbus, 1970.

Jones, A.J. : *Principles of Education in India*, McGraw Hill, New York, 1963.

Kak, Subhash C. : *Indian Physics: Outline of Early History*, ArXiv, London, 2003.

Khosbed, A.W. : *Education Movement in India*, NCERT, New Delhi, 1963.

Krause, M.V. and L.K. Mahann : *Food Nutrition and Diet Therapy*, W.B. Saunders Co., Philadelphia, 1979.

Krouse, H. : *Counselling Technique with Youth*, Charles and Mervil, Ohio, 1960.

Kumari, P. : *Role of Anganwadi Workers in Integrated Child Development Services Programme*, Discovery Publishing House, New Delhi, 1994.

Lagerstrom, B. : *Indian Women in Industry,* Zimbabwe House (Pvt) Ltd., New Delhi, 1984.

Maclean, M.S. : *Counselling of Technical Education,* McGraw Hill, New York, 1955.

Malcolm, S. : *Science and Technology for People,* Tamil Nader State Council for Science and Technology, Madras, 1985.

Marilyn, Carr : *Technology and Women: Conceptual and Empirical Issues,* Allen and Unwin, London, 2000.

Ndunda, M.M.K. : *Women's Resistance to Science Careers in India,* Queen's University, Kingston, 1990.

O'Kelly, Elizabeth : *Simple Technologies for Rural Women in Bangladesh and India,* Kuber Pub., Dhaka, 1978.

Oblsen, M.M. : *Technical Education,* Holt Rinehart and Winston, New York, 1977.

Pacey, A. : *Culture of Technology,* Cambridge, London, 1983.

Pinto, Diana V. : *Education : Theory, Research and Practice,* Rand McNally College Publishing Company, Chicago, 1978.

Potter, N.M. : *Food Science,* The AVI Publishing Co., Connecticut, 1978.

Rahman, A. : *Fatullah Shirazi, a Sixteenth-Century Indian Scientist,* MacMillan, London, 1968.

Rakowski, C.A. : *Women in Non-Traditional Industry: the Case of Steel in Ciudad Guayana,* Venezuela, London, 1985.

Samba, Murthy K.V. : *Science and Technology in Medieval India,* Pratap, Agra, 1982.

Sandler, J. : *The Tech and Tools Book: a Guide to Technologies Women are Using Worldwide,* International Women's Tribune Publications, New Delhi, 1986.

Sharma, P.V. : *History of Medicine in India,* Academic, New Delhi, 1992.

Singh, Harkishan : *Pharmaceutical Education,* Vallabh Prakashan, New Delhi, 1998.

Sponsors : *Proceedings of the Southeast Asian Seminar on Women and Science in Developing Countries,* Kovalevskaia Fund, New York, 1987.

Stadlmare, W.J. and. Cotterill, O.Y. : *Egg Science and Technology*, The AVI Publishing Co., Connecticut, 1973.

Stolte-Heiskanen, V. : *Science: Role of Changing India*, ISSC, New Delhi, 1991.

Subbarayappa, B.V. : *Science and Technological Exchanges between India and Soviet Central Asia*, Educational, New Delhi, 1985.

Theresa, Montini : *Science: Technology and Human Values*, Oxford, New York, 1993.

Tiwari, P.V. : *Kasyapa Samhita or Vrdhajivaka Tantra*, Visvabharati, Varanasi, 1997.

Tolber, E.L. : *Counselling for Career Development*, Houghton, Mifflin, 1974.

Uberoi, Chanchal : *India's Long History of Women in Science*, Physics Today, New Delhi, 1993.

Van, William N. : *Education in the 21st Century*, Houghton Mifflin Co., Boston, 2002.

Veken, M. Van Der : *Women: Technology and Development*, ATOL, Louvain, 1988.

Veldhnis, M.K. : *Food Science and Technology*, The AVI Publishing Co., Connecticut, 1977.

Vijaya Lakshmi, Sharma : *Science in Pre-Modern India*, D.K. Print World, New Delhi, 2001.

Virginia, Hazzard : *Village Technology and Women's Work in Eastern Africa*, UNICEF, Geneva, 1976.

Wardha, B. : *Science and Technology in India*, Centre of Science for Villages, Wardha, 1983.

Whitaker, J.R. : *Principles of Enzymology for Food Science*, Marcel Dekker, New York, 1977.

Winslow, A. : *Science: History*, Duke University Press, London, 1990.

Yukic, T.S. : *Fundamentals of Recreation*, Harper Row, London, 1963.

Zimmerman, J. : *Science in India: Today and Tomorrow*, Praeger Publishers, London, 1983.

Index

F

G

H

I

M

N

❑❑❑